FUZZY ARBITRARY
ORDER SYSTEM

FUZZY ARBITRARY ORDER SYSTEM

Fuzzy Fractional Differential Equations and Applications

SNEHASHISH CHAKRAVERTY

Department of Mathematics, National Institute of Technology Rourkela, Odisha, India

SMITA TAPASWINI

College of Mathematics and Statistics, Chongqing University, Chongqing, P.R. China
Department of Mathematics, KIIT University, Bhubaneswar, Odisha, India

DIPTIRANJAN BEHERA

Institute of Reliability Engineering, University of Electronic Science and Technology of China, Chengdu, Sichuan, P.R. China

Published by John Wiley & Sons, Inc., Hoboken, New Jersey
Published simultaneously in Canada

For general information on our other products and services or for technical support, please contact our Customer Care Department within the United States at (800) 762-2974, outside the United States at (317) 572-3993 or fax (317) 572-4002.

Wiley also publishes its books in a variety of electronic formats. Some content that appears in print may not be available in electronic formats. For more information about Wiley products, visit our web site at www.wiley.com.

Library of Congress Cataloging-in-Publication Data:

Names: Chakraverty, Snehashish. | Tapaswini, Smita, 1987- | Behera, D. (Diptiranjan), 1988-
Title: Fuzzy arbitrary order system : fuzzy fractional differential equations and applications / by Snehashish Chakraverty, Smita Tapaswini, D. Behera.
Description: Hoboken, New Jersey : John Wiley and Sons, Inc., [2016] | Includes bibliographical references and index.
Identifiers: LCCN 2016013567 (print) | LCCN 2016015086 (ebook) | ISBN 9781119004110 (cloth) | ISBN 9781119004134 (pdf) | ISBN 9781119004172 (epub)
Subjects: LCSH: Fractional differential equations. | Fuzzy mathematics. | Differential equations.
Classification: LCC QA314 .C43 2016 (print) | LCC QA314 (ebook) | DDC 515/.352–dc23
LC record available at http://lccn.loc.gov/2016013567

Typeset in 10/12pt TimesLTStd by SPi Global, Chennai, India

Printed in the United States of America
10 9 8 7 6 5 4 3 2 1

CONTENTS

PREFACE

Every physical problem is inherently biased by uncertainty. There is often a need to model, solve, and interpret the problems one encounters in the world of uncertainty. In general, science and engineering systems are governed by ordinary and partial differential equations, but the type of differential equation depends upon the application, domain, complicated environment, the effect of coupling, and so on. As such, the complicacy needs to be handled by recently developed arbitrary (fractional)-order differential equations. The arbitrary-order differential equations are themselves not easy to handle. In recent years, this subject has become an important area of research due to its wide range of applications in various disciplines, namely physics, chemistry, applied mathematics, biology, economics, and in engineering systems such as fluid mechanics, viscoelasticity, civil, mechanical, aerospace, and chemical. In general, parameters, variables, and initial conditions involved in the model are considered as crisp or defined exactly for easy computation. However, rather than the particular value, we may have only the vague, imprecise, and incomplete information about the variables and parameters being a result of errors in measurement, observations, experiment, applying different operating conditions, or it may be maintenance-induced errors, which are uncertain in nature. So, to overcome these uncertainties and vagueness, one may use either stochastic and statistical approach or interval and fuzzy set theory, but stochastic and statistical uncertainty occurs due to the natural randomness in the process. It is generally expressed by a probability density or frequency distribution function. For the estimation of the distribution, it requires sufficient information about the variables and parameters involved in it. On the other hand, interval and fuzzy set theory refers to the uncertainty when we may have lack of knowledge or incomplete information about the variables and parameters. As such, in this book, interval and fuzzy set theory has been used for the uncertainty analysis. These uncertainties are introduced in the general arbitrary (fractional)-order differential equations, which are named as Fuzzy Arbitrary-Order Differential Equations. Due to the complexity in the fuzzy arithmetic, one may need reliable and efficient analytical and numerical techniques for the solution of fuzzy arbitrary-order differential equations.

In view of the previous discussion, this book presents initially the basics of fuzzy and interval theory along with preliminaries of arbitrary (fractional)-order differential equations. Then various methods to solve fuzzy arbitrary (fractional)-order differential equations with fuzzy initial and/or boundary conditions are presented. The book consists of 14 chapters, and in order to understand the essence of fuzzy arbitrary-order differential equations, the developed methods have been applied then to solve various mathematical examples and application problems of engineering and sciences.

Accordingly, Chapter 1 addresses the preliminaries on fuzzy set theory, and Chapter 2 recalls the basics of fractional and fuzzy fractional differential equations. Chapter 3 deals with the analytical methods for the solution of n-term fuzzy fractional differential equations. The concept of n-term fuzzy fractional linear differential equations is briefly discussed here. As the sign of the coefficients in the fuzzy fractional-order differential equations plays a very important role, three possible cases, namely when all the coefficients are positive, when all the coefficients are negative, and when the coefficients are combinations of positive and negative, are all discussed. Methods based on fuzzy center, addition and subtraction of fuzzy numbers, and double parametric form of fuzzy numbers are also included here. In Chapter 4, numerical schemes, namely homotopy perturbation method (HPM), Adomian decomposition method (ADM), and variational iteration method (VIM) have been presented for fuzzy fractional differential equations. Chapters 3 and 4 also contain simple mathematical examples for better understanding of these methods. Solution of fuzzy arbitrary-order heat equations using HPM has been addressed in Chapter 5. Fuzziness in the initial conditions is taken in terms of triangular fuzzy number. Chapter 6 presents the solution of fuzzy arbitrary-order predator–prey equation. In the predator–prey equation, fuzziness in the initial conditions, which is again taken in the form of triangular fuzzy number and solution, is obtained by HPM. Comparisons are also made with crisp solutions. Numerical solution of uncertain arbitrary-order Rossler's system has been analyzed in Chapter 7. It is worth mentioning that Rossler's system was found to be useful in the modeling of equilibrium in chemical reactions. Chapter 8 describes the numerical solution of imprecisely defined fractionally damped structural systems. In this regard, both discrete and continuous systems have been taken into consideration subjected to unit impulse and step loads. First, a mechanical spring–mass system having fractional damping of order 1/2 with fuzzy initial condition has been taken to analyze the discrete system. Fuzziness in the initial conditions is modeled through different types of convex, normalized fuzzy sets, namely triangular, trapezoidal, and Gaussian fuzzy numbers. HPM is used with fuzzy-based approach to obtain the uncertain impulse response. Next, this chapter includes the study of fuzzy fractionally damped continuous system that is a beam using the double parametric form of fuzzy numbers subject to unit step and impulse loads. HPM is used for obtaining the fuzzy response and various numerical examples are solved. Chapter 9 gives the double parametric form of fuzzy numbers to solve fuzzy fractional diffusion equation with initial conditions as triangular and Gaussian fuzzy numbers. In the solution process, HPM and ADM are used. Lastly, Chapter 10 presents a type of traveling-wave problem, namely the nonlinear interval fractional Fornberg–Whitham equations subject to interval initial conditions. VIM has been applied to obtain the uncertain solution.

Further, double-parametric-based method has also been used to solve various other fuzzy fractional differential equations, namely vibration equation of large membrane, telegraph equation, and Fokker–Planck equation in Chapters 11–13, respectively. Finally, Chapter 14 addresses the solution of fuzzy fractional Bagley–Torvik equations using HPM following concepts of Hukuhara derivative. It is worth mentioning that an appendix has also been included for handling crisp differential equations with fractionally damped spring–mass and beam problems for the sake of completeness so that readers may go through the same to have ready reference of crisp cases.

This book aims to provide basic concepts of fuzzy arbitrary (/fractional)-order differential equations with various important applications in science and engineering systems in a systematic manner along with the recent trends, usefulness, and developments. The book will certainly find an important source for graduate and postgraduate students, teachers, and researchers in colleges, universities/institutes, and industries in various science and engineering fields, wherever one wants to model and analyze their uncertain physical problems. It is known that uncertainty is a must in every field of science and engineering, so this work will prove to be a handy and important book to handle their problems.

Finally, we do believe that the book may represent a new vista because it demonstrates how the most current, advanced, and revolutionary mathematical and computational techniques can be put to effective use of fuzzy and interval analysis in the uncertain arbitrary-order differential equations.

S. Chakraverty, Rourkela
Smita Tapaswini, Chongqing
Diptiranjan Behera, Chengdu

ACKNOWLEDGMENTS

Writing this book was an amazing journey that would not have been possible without the continuous support, encouragement, and motivation received from many outstanding people around us who not only guided us through our hardships, but also made us believe that we could achieve what we wanted to.

As such, the first author would firstly like to thank his parents for being his pillars of motivation. Next, he would like to thank his wife, Mrs. Shewli Chakraborty, for her immense love and support. He would also like to thank his daughters, Shreyati and Susprihaa, for their love and source of inspiration during the course of writing this book. The support of all the Ph.D. students of the first author and the NIT Rourkela facilities are also gratefully acknowledged.

The second author would like to thank her family members for their continuous source of encouragement and motivation. Writing this book would not have been possible without the love, concern, and motivation of her father Mr. Kedarnath Behera, mother Mrs. Sandhyarani Behera, and brother Mr. Deepak Behera.

The third author would like to express his sincere gratitude to his parents, Mr. Muralidhar Behera and Mrs. Sachala Behera, and elder brothers, Manoranjan and Srutiranjan, for their unwavering support and invariable source of motivation. He firmly believes that without their support he would not have been able to fly high and achieve success.

The second and third authors, namely Dr. Diptiranjan Behera and Dr. Smita Tapaswini of this book, are highly obliged to the family members of the first author, especially to his wife Mrs. Shewli Chakraborty and daughters Shreyati and Susprihaa for their continuous love, support, and source of inspiration at all the time.

We are grateful to all the contributors and to the authors of the books and journal/conference papers listed in the book for providing us with valuable information. Finally, we heartily acknowledge the support and help of the editorial team of the publisher throughout this project to complete it in the targeted time.

S. Chakraverty
Smita Tapaswini
Diptiranjan Behera

1

PRELIMINARIES OF FUZZY SET THEORY

This chapter presents the notations, definitions of fuzzy numbers (namely triangular, trapezoidal, Gaussian, double parametric form), type of differentiability, theorems/lemma related to fuzzy/fuzzy fractional differential equations, and fuzzy/interval arithmetic, which are relevant to the current investigation. Several excellent books related to this have also been written by different authors representing the scope and various aspects of fuzzy set theory such as in Zimmermann (2001), Jaulin et al. (2001), Ross (2004), Hanss (2005), Moore (1966), and Chakraverty (2014). These books also give an extensive review on fuzzy set theory and its applications, which may help the reader in understating the basic concepts of fuzzy set theory and its application.

Definition 1.1 Interval An interval $\tilde{x}$ is denoted by $[\underline{x}, \overline{x}]$ on the set of real numbers R given by

$$\tilde{x} = [\underline{x}, \overline{x}] = \{x \in R : \underline{x} \le x \le \overline{x}\}. \tag{1.1}$$

We have only considered closed intervals throughout this thesis, although there exist various other types of intervals such as open and half-open intervals. $\underline{x}$ and $\overline{x}$ are known as the left and right end points, respectively, of the interval $\tilde{x}$ in Eq. (1.1).

Fuzzy Arbitrary Order System: Fuzzy Fractional Differential Equations and Applications, First Edition.
Snehashish Chakraverty, Smita Tapaswini, and Diptiranjan Behera.
© 2016 John Wiley & Sons, Inc. Published 2016 by John Wiley & Sons, Inc.

Let us now consider two arbitrary intervals $\widetilde{x} = [\underline{x}, \overline{x}]$ and $\widetilde{y} = [\underline{y}, \overline{y}]$. These two intervals are said to be equal if they are in the same set. Mathematically, it only happens when corresponding end points are equal. Hence, one may write

$$\widetilde{x} = \widetilde{y} \quad \text{if and only if} \quad \underline{x} = \underline{y} \text{ and } \overline{x} = \overline{y}. \tag{1.2}$$

For the given two arbitrary intervals $\widetilde{x} = [\underline{x}, \overline{x}]$ and $\widetilde{y} = [\underline{y}, \overline{y}]$, interval arithmetic operations such as addition $(+)$, subtraction $(-)$, multiplication $(\times)$, and division $(/)$ are defined as follows:

$$\widetilde{x} + \widetilde{y} = [\underline{x} + \underline{y}, \overline{x} + \overline{y}], \tag{1.3}$$

$$\widetilde{x} - \widetilde{y} = [\underline{x} - \overline{y}, \overline{x} - \underline{y}], \tag{1.4}$$

$$\widetilde{x} \times \widetilde{y} = [\min S, \max S], \quad \text{where } S = \{\underline{x} \times \underline{y}, \underline{x} \times \overline{y}, \overline{x} \times \underline{y}, \overline{x} \times \overline{y}\}, \tag{1.5}$$

and

$$\widetilde{x}/\widetilde{y} = [\underline{x}, \overline{x}] \times \left[\frac{1}{\overline{y}}, \frac{1}{\underline{y}}\right] \quad \text{if } 0 \notin \widetilde{y}. \tag{1.6}$$

Now if k is a real number and $\widetilde{x} = [\underline{x}, \overline{x}]$ is an interval, then the multiplication of them is given by

$$k\widetilde{x} = \begin{cases} [k\overline{x}, k\underline{x}], & k < 0, \\ [k\underline{x}, k\overline{x}], & k \geq 0. \end{cases} \tag{1.7}$$

Definition 1.2 Fuzzy Number A fuzzy number $\widetilde{U}$ is convex, normalized fuzzy set $\widetilde{U}$ of the real line R such that

$$\{\mu_{\widetilde{U}}(x) : R \to [0, 1], \quad \forall x \in R\},$$

where, $\mu_{\widetilde{U}}$ is called the membership function of the fuzzy set, and it is piecewise continuous. There exists a variety of fuzzy numbers. But in this study, we have used only the triangular, trapezoidal, and Gaussian fuzzy numbers. So, we define these three fuzzy numbers as follows.

Definition 1.3 Triangular Fuzzy Number (TFN) A triangular fuzzy number (TFN) $\widetilde{U}$ is a convex, normalized fuzzy set $\widetilde{U}$ of the real line R such that

1. There exists exactly one $x_0 \in R$ with $\mu_{\widetilde{U}}(x_0) = 1$ (x_0 is called the mean value of $\widetilde{U}$), where $\mu_{\widetilde{U}}$ is called the membership function of the fuzzy set.
2. $\mu_{\widetilde{U}}(x)$ is piecewise continuous.

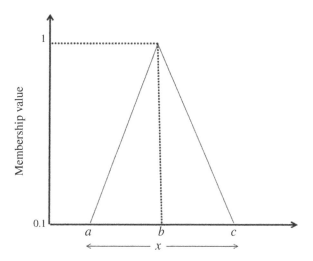

Figure 1.1 Triangular fuzzy number

Let us consider an arbitrary TFN $\tilde{U} = (a, b, c)$. The membership function $\mu_{\tilde{U}}$ of $\tilde{U}$ is defined as follows:

$$\mu_{\tilde{U}}(x) = \begin{cases} 0, & x \le a, \\ \dfrac{x-a}{b-a}, & a \le x \le b \\ \dfrac{c-x}{c-b}, & b \le x \le c \\ 0, & x \ge c. \end{cases}$$

The TFN $\tilde{U} = (a, b, c)$ can be represented by an ordered pair of functions through r-cut approach, namely $[\underline{u}(r), \overline{u}(r)] = [(b-a)r + a, -(c-b)r + c]$ where, $r \in [0, 1]$ (Fig. 1.1).

Definition 1.4 Trapezoidal Fuzzy Number (TrFN) We consider an arbitrary trapezoidal fuzzy number (TrFN) $\tilde{U} = (a, b, c, d)$. The membership function $\mu_{\tilde{U}}$ of $\tilde{U}$ is given as

$$\mu_{\tilde{U}}(x) = \begin{cases} 0, & x \le a \\ \dfrac{x-a}{b-a}, & a \le x \le b \\ 1, & b \le x \le c \\ \dfrac{d-x}{d-c}, & c \le x \le d \\ 0, & x \ge d. \end{cases}$$

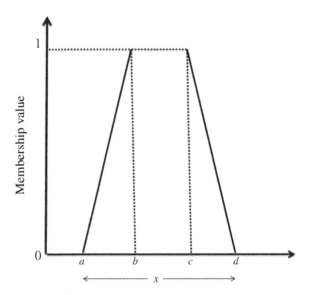

Figure 1.2 Trapezoidal fuzzy number

The TrFN $\tilde{U} = (a,\ b,\ c, d)$ can be represented with an ordered pair of functions through r-cut approach that is $[\underline{u}(r),\ \overline{u}(r)] = [(b-a)r+a,\ -(d-c)r+d]$ where, $r \in [0,\ 1]$ (Fig. 1.2).

Definition 1.5 Gaussian Fuzzy Number (GFN) Let us now define an arbitrary asymmetrical Gaussian fuzzy number, $\tilde{U} = (\delta, \sigma_l, \sigma_r)$. The membership function $\mu_{\tilde{U}}$ of $\tilde{U}$ will be as follows:

$$\mu_U(x) = \begin{cases} \exp\left[\dfrac{-(x-\delta)^2}{2\sigma_l^2}\right] & \text{for } x \le \delta \\[4mm] \exp\left[\dfrac{-(x-\delta)^2}{2\sigma_r^2}\right] & \text{for } x \ge \delta \end{cases} \qquad \forall x \in R,$$

where, the modal value is denoted as δ and σ_l, σ_r denote the left-hand and right-hand spreads (fuzziness), respectively, corresponding to the Gaussian distribution. For symmetric Gaussian fuzzy number, the left-hand and right-hand spreads are equal, that is, $\sigma_l = \sigma_r = \sigma$. So the symmetric Gaussian fuzzy number may be written as $\tilde{U} = (\delta, \sigma, \sigma)$ and the corresponding membership function may be defined as $\mu_{\tilde{U}}(x) = \exp\{-\beta(x-\delta)^2\}\ \forall x \in R$ where $\lambda = 1/2\sigma^2$. The symmetric Gaussian fuzzy number (Fig. 1.3) in parametric can be represented as

$$\tilde{U} = [\underline{u}(r),\ \overline{u}(r)] = \left[\delta - \sqrt{-\frac{(\log_e r)}{\lambda}},\ \delta + \sqrt{-\frac{(\log_e r)}{\lambda}}\right], \quad \text{where } r \in [0,\ 1].$$

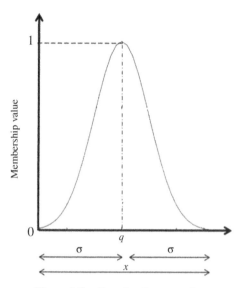

Figure 1.3 Gaussian fuzzy number

For all the aforementioned type of fuzzy numbers, the lower and upper bounds of the fuzzy numbers satisfy the following requirements:

(i) $\underline{u}(r)$ is a bounded left-continuous nondecreasing function over $[0, 1]$;

(ii) $\overline{u}(r)$ is a bounded right-continuous nonincreasing function over $[0, 1]$;

(iii) $\underline{u}(r) \leq \overline{u}(r), 0 \leq r \leq 1$.

Definition 1.6 Double Parametric Form of Fuzzy Number Using the r-cut approach as discussed in Definitions 1.2–1.5 for all the fuzzy numbers, we have $U = [\underline{u}(r), \overline{u}(r)]$. Now one may write this as crisp number with double parametric form as $\tilde{U}(r, \beta) = \beta(\overline{u}(r) - \underline{u}(r)) + \underline{u}(r)$ where r and $\beta \in [0, 1]$. To obtain the lower and upper bounds of the solution in single parametric form, we may use $\beta = 0$ and 1, respectively. This may be represented as $\tilde{U}(r, 0) = \underline{u}(r)$ and $\tilde{U}(r, 1) = \overline{u}(r)$.

Definition 1.7 Fuzzy Center Fuzzy center of an arbitrary fuzzy number $\tilde{u} = [\underline{u}(r), \overline{u}(r)]$ is defined as $\tilde{u}^c = \frac{\underline{u}(r) + \overline{u}(r)}{2}$, for all $0 \leq r \leq 1$.

Definition 1.8 Fuzzy Radius Fuzzy radius of an arbitrary fuzzy number $\tilde{u} = [\underline{u}(r), \overline{u}(r)]$ is defined as $\Delta\tilde{u} = \frac{\overline{u}(r) - \underline{u}(r)}{2}$ for all $0 \leq r \leq 1$.

Definition 1.9 Fuzzy Width Fuzzy space or width of an arbitrary fuzzy number $\tilde{u} = [\underline{u}(r), \overline{u}(r)]$ is defined as $|\underline{u}(r) - \overline{u}(r)|$, for all $0 \leq r \leq 1$.

Definition 1.10 Fuzzy Arithmetic For any two arbitrary fuzzy numbers $\tilde{x} = [\underline{x}(r), \overline{x}(r)]$, $\tilde{y} = [\underline{y}(r), \overline{y}(r)]$ and scalar k, the fuzzy arithmetic is similar to the interval arithmetic defined as follows:

(i) $\tilde{x} = \tilde{y}$ if and only if $\underline{x}(r) = \underline{y}(r)$ and $\overline{x}(r) = \overline{y}(r)$

(ii) $\tilde{x} + \tilde{y} = [\underline{x}(r) + \underline{y}(r), \overline{x}(r) + \overline{y}(r)]$

(iii) $\tilde{x} \times \tilde{y} = [\min(S), \max(S)]$ where

$$S = \{\underline{x}(r) \times \underline{y}(r), \underline{x}(r) \times \overline{y}(r), \overline{x}(r) \times \underline{y}(r), \overline{x}(r) \times \overline{y}(r)\}$$

(iv) $k\tilde{x} = \begin{cases} [k\overline{x}(r), k\underline{x}(r)], & k < 0 \\ [k\underline{x}(r), k\overline{x}(r)], & k \geq 0 \end{cases}$

(v) $\dfrac{\tilde{x}}{\tilde{y}} = [\underline{x}(r), \overline{x}(r)] \times \left[\dfrac{1}{\overline{y}(r)}, \dfrac{1}{\underline{y}(r)}\right]$, where $0 \notin \tilde{y}$, where $0 \notin \tilde{y}$.

Definition 1.11 Let $F : (a, b) \to R_F$ and $t_0 = (a, b)$ (Khastan et al., 2011; Chalco-Cano and Roman-Flores, 2008). X is called differentiable at t_0, if there exists $F'(t_0) \in R_F$ such that

(i) for all $h > 0$ sufficiently close to 0, the Hukuhara difference $F(t_0 + h) \ominus F(t_0)$ and $F(t_0) \ominus F(t_0 - h)$ exists and (in metric D)

$$\lim_{h \to 0^+} \frac{F(t_0 + h) \ominus F(t_0)}{h} = \lim_{h \to 0^+} \frac{F(t_0) \ominus F(t_0 - h)}{h} = F'(t_0),$$

or

(ii) for all $h > 0$ sufficiently close to 0, the Hukuhara difference $F(t_0) \ominus F(t_0 + h)$ and $F(t_0 - h) \ominus F(t_0)$ exists and (in metric D)

$$\lim_{h \to 0^+} \frac{F(t_0) \ominus F(t_0 + h)}{-h} = \lim_{h \to 0^+} \frac{F(t_0 - h) \ominus F(t_0)}{-h} = F'(t_0)$$

Chalco-Cano and Roman-Flores (2008) used Definition 1.11 to obtain the following results.

Theorem 1.1 Let $F : (a, b) \to R_F$ and denote $[F(t; r)] = [\underline{f}(t; r), \overline{f}(t; r)]$ for each $r \in [0, 1]$.

(i) If F is differentiable of the first type (I), then $\underline{f}(t; r)$ and $\overline{f}(t; r)$ are differentiable functions, and we have $[F'(t; r)] = [\underline{f}'(t; r), \overline{f}'(t; r)]$.

(ii) If F is differentiable of the second type (II), then $\underline{f}(t; r)$ and $\overline{f}(t; r)$ are differentiable functions, and we have $[F'(t; r)] = [\overline{f}'(t; r), \underline{f}'(t; r)]$.

Proof The proof of the theorem is given in Chalco-Cano and Roman-Flores (2008).

BIBLIOGRAPHY

Chakraverty S. *Mathematics of Uncertainty Modeling in the Analysis of Engineering and Science Problems*. USA: IGI Global Publication; 2014.

Chalco-Cano Y, Roman-Flores H. On new solutions of fuzzy differential equations. *Chaos Solitons Fract* 2008;**38**:112–119.

Hanss M. *Applied Fuzzy Arithmetic: An Introduction with Engineering Applications*. Berlin: Springer-Verlag; 2005.

Jaulin L, Kieffer M, Didri OT, Walter E. *Applied Interval Analysis*. London: Springer; 2001.

Khastan A, Nieto JJ, Rodriguez-Lopez R. Variation of constant formula for first order fuzzy differential equations. *Fuzzy Sets Syst* 2011;**177**:20–33.

Moore RE. *Interval Analysis*. Englewood Cliffs: Prentice Hall; 1966.

Ross TJ. *Fuzzy Logic with Engineering Applications*. New York: John Wiley & Sons; 2004.

Zimmermann HJ. *Fuzzy Set Theory and Its Application*. Boston/Dordrecht/London: Kluwer Academic Publishers; 2001.

2

BASICS OF FRACTIONAL AND FUZZY FRACTIONAL DIFFERENTIAL EQUATIONS

We discussed in the previous chapter about fuzzy set theory and related definitions and notations, which will be helpful for understanding the fuzzy and fuzzy fractional differential equations. In this chapter, we present some preliminaries related to fuzzy differential equations and fractional and fuzzy fractional differential equations, which will be used further in this book.

Definition 2.1 Fuzzy Initial Value Problem (FIVP) Let us consider the nth order fuzzy initial value problem (FIVP)

$$\widetilde{y}^{(n)}(t) + a_{n-1}(t)\widetilde{y}^{(n-1)}(t) + \cdots + a_1(t)\widetilde{y}'(t) + a_0(t)\widetilde{y}(t) = \widetilde{g}(t), \qquad (2.1)$$

where $\widetilde{y}$ is a fuzzy function of t, $\widetilde{y}^{(n)}(t), \widetilde{y}^{(n-1)}(t), \ldots, \widetilde{y}'(t), \widetilde{y}(t)$ are the Hukahara fuzzy derivatives, $a_i(x), 0 \le i \le n - 1$, continuous on some interval I, subject to fuzzy initial conditions

$$\widetilde{y}(0) = \widetilde{b}_0, \widetilde{y}'(0) = \widetilde{b}_1, \ldots, \widetilde{y}^{(n-1)}(0) = \widetilde{b}_{n-1}.$$

Definition 2.2 Fuzzy Boundary Value Problem (FBVP) Accordingly, let us consider the n-th order FBVP as

$$\widetilde{y}^{(n)}(t; r) + a_{n-1}(t)\widetilde{y}^{(n-1)}(t; r) + \cdots + a_1(t)\widetilde{y}'(t; r) + a_0(t)\widetilde{y}(t; r) = \widetilde{g}(t; r), \qquad (2.2)$$

Fuzzy Arbitrary Order System: Fuzzy Fractional Differential Equations and Applications, First Edition.
Snehashish Chakraverty, Smita Tapaswini, and Diptiranjan Behera.
© 2016 John Wiley & Sons, Inc. Published 2016 by John Wiley & Sons, Inc.

where $a_i(t), 0 \leq i \leq n - 1$, is continuous on some interval I, subject to fuzzy boundary conditions

$$\widetilde{y}(a; r) = \left[\underline{\beta}(r), \overline{\beta}(r) \right], \quad \widetilde{y}(b; r) = \left[\underline{\gamma}(r), \overline{\gamma}(r) \right],$$

and $\widetilde{y}(t; r)$ is the solution to be determined.

Lemma 2.1   If $\widetilde{u}(t) = (x(t), y(t), z(t))$ is a fuzzy triangular number valued function and if $\widetilde{u}$ is Hukuhara differentiable (Bede, 2008), then $\widetilde{u}' = (x', y', z')$.

By using Hukuhara differentiable, we intend to solve the FIVP

$$\widetilde{x}' = f(t, \widetilde{x}), \tag{2.3}$$

subject to triangular fuzzy initial condition

$$\widetilde{x}(t_0) = \widetilde{x}_0,$$

where, $\widetilde{x}_0 = (\underline{x}_0, x_0^c, \overline{x}_0) \in R, \widetilde{x}(t) = (\underline{u}, u^c, \overline{u}) \in R$, and

$$f : \left[t_0, t_0 + a \right] \times R \to R, f\left(t, \left(\underline{u}, u^c, \overline{u} \right) \right)$$

$$= \left(\underline{f}\left(t, \underline{u}, u^c, \overline{u} \right), f^c\left(t, \underline{u}, u^c, \overline{u} \right), \overline{f}\left(t, \underline{u}, u^c, \overline{u} \right) \right).$$

We can translate this into the following system of ordinary differential equations:

$$\begin{cases} \underline{u} = \underline{f}(t, \underline{u}, u^c, \overline{u}) \\ u^c = f^c(t, \underline{u}, u^c, \overline{u}) \\ \overline{u} = \overline{f}(t, \underline{u}, u^c, \overline{u}) \\ \underline{u}(0) = \underline{x}_0, u^c(0) = x_0^c, \overline{u}(0) = \overline{x}_0. \end{cases} \tag{2.4}$$

Definition 2.3 Riemann–Liouville Fractional Integral (Podlubny, 1999) The Riemann–Liouville integral operator J^α of order $\alpha > 0$ is defined as

$$J^\alpha f(t) = \frac{1}{\Gamma(\alpha)} \int_0^t (t - \tau)^{\alpha-1} f(\tau) d\tau, \quad t > 0. \tag{2.5}$$

Definition 2.4 Fuzzy Riemann–Liouville Fractional Integral (Mazandarani and Kamyad, 2013) The Riemann–Liouville fractional integral of order α of the fuzzy number valued function f, based on its r-cut representations, can be expressed as

$$\left[J^\alpha \widetilde{f}(t; r) \right] = \left[J^\alpha \underline{f}(t; r), J^\alpha \overline{f}(t; r) \right], \quad t > 0,$$

where

$$J^{\alpha}\underline{f}(t) = \frac{1}{\Gamma(\alpha)} \int_0^t (t-\tau)^{\alpha-1} \underline{f}(\tau)d\tau, \quad t > 0,$$

$$J^{\alpha}\overline{f}(t) = \frac{1}{\Gamma(\alpha)} \int_0^t (t-\tau)^{\alpha-1} \overline{f}(\tau)d\tau, \quad t > 0.$$

Definition 2.5 Caputo Derivative (Podlubny, 1999) The fractional derivative of $f(t)$ in the Caputo sense is defined as follows:

$$D^{\alpha}f(t) = J^{m-\alpha}D^m f(t) = \begin{cases} \dfrac{1}{\Gamma(m-\alpha)} \displaystyle\int_0^t \dfrac{f^{(m)}(\tau)d\tau}{(t-\tau)^{\alpha+1-m}}, & m-1 < \alpha < m, \quad m \in N \\[4mm] \dfrac{d^m}{dt^m}f(t), & \alpha = m, \quad m \in N. \end{cases}$$

$$(2.6)$$

Some basic properties of the fractional operator are as follows:

(i) $J^{\alpha}J^{\gamma}f(t) = J^{\alpha+\gamma}f(t), \ \alpha, \gamma \geq 0$

(ii) $J^{\alpha}(t^{\gamma}) = \begin{cases} \frac{\Gamma(\gamma+1)t^{\alpha+\gamma}}{\Gamma(\alpha+\gamma+1)} \end{cases}, \ \alpha > 0, \ \gamma > -1, \ t > 0.$

Definition 2.6 Caputo-Type Fuzzy Fractional Derivatives (Mazandarani and Kamyad, 2013) Let $\widetilde{f}(t; r)$ be a fuzzy valued function and $[\widetilde{f}(t; r)] = [\underline{f}(t; r), \overline{f}(t; r)]$, for $r \in [0, 1]$, $0 < \alpha < 1$, and $t \in (a, b)$.

(a) If $\widetilde{f}(t; r)$ is a Caputo-type fuzzy fractional differentiable function in the first form, then
$$\left[D^{\alpha}\widetilde{f}(t; r) \right] = \left[D^{\alpha}\underline{f}(t; r), D^{\alpha}\overline{f}(t; r) \right].$$

(b) If $\widetilde{f}(t; r)$ is a Caputo-type fuzzy fractional differentiable function in the second form, then
$$\left[D^{\alpha}\widetilde{f}(t; r) \right] = \left[D^{\alpha}\overline{f}(t; r), D^{\alpha}\underline{f}(t; r) \right].$$

where,

$$D^{\alpha}\underline{f}(t; r) = \frac{1}{\Gamma(m-\alpha)} \int_0^t \frac{\underline{f}^{(m)}(\tau)d\tau}{(t-\tau)^{\alpha+1-m}}, \quad m-1 < \alpha < m, \quad m \in N,$$

$$D^\alpha \overline{f}(t; r) = \frac{1}{\Gamma(m - \alpha)} \int_0^t \frac{\overline{f}^{(m)}(\tau) d\tau}{(t - \tau)^{\alpha + 1 - m}}, \quad m - 1 < \alpha < m, \quad m \in N,$$

$$\frac{d^m}{dt^m} f(t), \quad \alpha = m, \quad m \in N.$$

Definition 2.7 Fractional Initial Value Problem (Podlubny, 1999) Let us consider the following fractional initial value problem (FrIVP):

$$D^\alpha y(t) = f(t, y), \tag{2.7}$$

subject to the initial condition

$$y(0) = y_0, \, t \in [a, b], \, \alpha \in (0, 1),$$

where D^α denotes the Caputo fractional differential operator.

Next, we combine differential equations of fractional order and with uncertainty, to consider a new type of dynamical system, that is, fuzzy differential equations of fractional order.

Definition 2.8 Fuzzy Fractional Initial Value Problem (Mazandarani and Kamyad, 2013) Let us consider the following fuzzy fractional initial value problem (FFIVP):

$$D^\alpha \widetilde{y}(t) = f(t, \widetilde{y}), \tag{2.8}$$

subject to the fuzzy initial condition

$$\widetilde{y}(0) = \widetilde{y}_0, \quad t \in [a, b], \quad \alpha \in (0, 1).$$

The FFIVP (2.8) can be considered equivalent by the following initial value problems:

$$\left[D^\alpha \underline{y}(t), D^\alpha \overline{y}(t) \right] = \left[\underline{f}(t, y), \overline{f}(t, \overline{y}) \right],$$

subject to the fuzzy initial condition

$$\left[\underline{y}(0), \overline{y}(0) \right] = \left[\underline{y}_0, \overline{y}_0 \right].$$

BIBLIOGRAPHY

Agarwal RP, Lakshmikantham V, Nieto JJ. On the concept of solution for fractional differential equations with uncertainty. *Nonlinear Anal* 2010;**72**:2859–2862.

Allahviranloo T, Salahshour S, Abbasbandy S. Explicit solutions of fractional differential equations with uncertainty. *Soft Comput* 2012;**16**:297–302.

Arshad S, Lupulescu V. On the fractional differential equations with uncertainty. *Nonlinear Anal* 2011;**74**:3685–3693.

Bede B. Note on numerical solutions of fuzzy differential equations by predictor-corrector method. *Inform Sci* 2008;**178**:1917–1922.

Diamond P, Kloeden P. *Metric Spaces of Fuzzy Sets: Theory and applications.* Singapore: World Scientific; 1994.

Kilbas AA, Srivastava HM, Trujillo JJ. *Theory and Application of Fractional Differential Equations.* Amsterdam: Elsevier Science B.V.; 2006.

Kiryakova VS. *Generalized Fractional Calculus and Applications.* England: Longman Scientific and Technical; 1993.

Lakshmikantham V, Mohapatra RN. *Theory of Fuzzy Differential Equations and Applications.* London: Taylor & Francis; 2003.

Mazandarania M, Kamyad VA. *Modified fractional Euler method for solving fuzzy fractional initial value problem* 2013;**18**:12–21.

Miller KS, Ross B. *An Introduction to the Fractional Calculus and Fractional Differential Equations.* New York: John Wiley & Sons; 1993.

Nieto JJ, Rodriguez-Lopez FD. Linear first-order fuzzy differential equation. *Int J Uncertain Fuzz Knowl.-Based Syst* 2006;**14**:687–709.

Oldham KB, Spanier J. *The Fractional Calculus.* New York, NY: Academic Press; 1974.

Podlubny I. *Fractional Differential Equations.* New York, NY: Academic Press; 1999.

Prakash P, Nieto JJ, Senthilvelavan S, Sudha Priya G. Fuzzy fractional initial value problem. *J Intell Fuzzy Syst* 2015;**28**:2691–2704.

Ross TJ. *Fuzzy Logic with Engineering Applications.* New York: John Wiley & Sons; 2004.

Salahshour S, Allahviranloo T, Abbasbandy S. Solving fuzzy fractional differential equations by fuzzy Laplace transforms. *Commun Nonlinear Sci Numer Simul* 2012;**17**:1372–1381.

Samko SG, Kilbas AA, Marichev OI. *Fractional Integrals and Derivatives: Theory and Applications.* Langhorne, PA: Gordon and Breach Science Publishers; 1993.

Tapaswini S, Chakraverty S. New analytical method for solving *n*-th order fuzzy differential equations. *Ann Fuzzy Math Inform* 2014;**8**:231–244.

Zimmermann HJ. *Fuzzy Set Theory and its Application.* London: Kluwer Academic Publishers; 2001.

3

ANALYTICAL METHODS FOR FUZZY FRACTIONAL DIFFERENTIAL EQUATIONS (FFDES)

Several numerical and/or analytical methods are available to solve fuzzy fractional differential equations (FFDEs). Analytical solutions play a significant role in proper understanding of qualitative features of various science and engineering problems. But it is not always possible to obtain the analytical solution of the same. We have discussed in the introduction that few researchers have developed analytical methods to obtain solution of n-term FFDEs. But it has been seen that the existing methods always convert the FFDEs to two crisp differential equations (Arshad and Lupulescu, 2011; Mohammed et al., 2011; Prakash et al., 2015; Salah et al., 2013) or sometime coupled differential equations, depending upon the sign of the coefficients. Accordingly, more computational time is needed to solve the coupled or system of FFDEs. This motivates the study to develop new analytical methods with less computation time. In this regard, first, we will briefly discuss the concept of n-term fuzzy fractional linear ordinary differential equations. In general, three cases may arise according to sign of the coefficients in the FFDEs. Finally, this chapter includes three analytical methods by considering three cases to solve n-term linear FFDEs as follows.

Method 1: Fuzzy-Center-Based Method (FCM) (Tapaswini and Chakraverty, 2014a).

Method 2: Method Based on Addition and Subtraction of Fuzzy Numbers (ASFM).

Method 3: Double-Parametric-Based Method (DPM) (Tapaswini and Chakraverty, 2013, 2014a).

Fuzzy Arbitrary Order System: Fuzzy Fractional Differential Equations and Applications, First Edition. Snehashish Chakraverty, Smita Tapaswini, and Diptiranjan Behera.
© 2016 John Wiley & Sons, Inc. Published 2016 by John Wiley & Sons, Inc.

First, we discuss the general form of n-term linear FFDEs along with the possible three cases in detail.

3.1 n-TERM LINEAR FUZZY FRACTIONAL LINEAR DIFFERENTIAL EQUATIONS

Let us consider n-term FFDEs in general form as

$$D^{\lambda_n}\widetilde{y}(t;r) + a_{n-1}(t)D^{\lambda_{n-1}}\widetilde{y}(t;r) + \cdots + a_1(t)D^{\lambda_1}\widetilde{y}(t;r) + a_0(t)D^{\lambda_0}\widetilde{y}(t;r) = \widetilde{g}(t;r),$$
$$(3.1)$$

subject to fuzzy initial conditions

$$\widetilde{y}^{(k)}(0;r) = \widetilde{b}_k(r), \quad k = 0, 1, 2, \ldots, n-1, \quad n-1 \le \lambda < n,$$

where $n+1 > \lambda \ge n > \lambda_{n-1} > \cdots > \lambda_1 > \lambda_0$ and $a_i(t), 0 \le i \le n-1$ are real constants, $\widetilde{b}_k, 0 \le k \le n-1$ are fuzzy numbers, and D^{λ_n} denotes the fractional derivative in Caputo sense. Here, $\widetilde{y}(t;r)$ is the solution to be determined.

Through r-cut approach, we may write the given fuzzy differential equation (Eq. (3.1)) as

$$\left[D^{\lambda_n}\underline{y}(t;r), D^{\lambda_n}\overline{y}(t;r)\right] + a_{n-1}(t)\left[D^{\lambda_{n-1}}\underline{y}(t;r), D^{\lambda_{n-1}}\overline{y}(t;r)\right] + \cdots$$
$$+ a_1\left[D^{\lambda_1}\underline{y}(t;r), D^{\lambda_1}\overline{y}(t;r)\right] + a_0(t)\left[D^{\lambda_0}\underline{y}(t;r), D^{\lambda_0}\overline{y}(t;r)\right] = \left[\underline{g}(t;r), \overline{g}(t;r)\right],$$
$$(3.2)$$

subject to fuzzy initial condition

$$\left[\underline{y}^{(k)}(0;r), \overline{y}^{(k)}(0;r)\right] = \left[\underline{b}_k(r), \overline{b}_k(r)\right], \quad k = 0, 1, 2, \ldots, n-1,$$
$$n-1 \le \lambda < n, \quad \text{where } r \in [0,1].$$

One may note that we have three cases arising with respect to the sign of the coefficients.

As such, we discuss these three cases in the following.

Case 1   Coefficients $a_{n-1}(t), a_{n-2}(t), \ldots, a_1(t), a_0(t)$ are all positive. By using the definition of Hukuhara derivative (Bede, 2008), one may write Eq. (3.2) as

$$D^{\lambda_n}\underline{y}(t;r) + a_{n-1}(t)D^{\lambda_{n-1}}\underline{y}(t;r) + \cdots + a_1 D^{\lambda_1}\underline{y}(t;r) + a_0(t)D^{\lambda_0}\underline{y}(t;r) = \underline{g}(t;r),$$
$$(3.3)$$

and

$$D^{\lambda_n}\overline{y}(t;r) + a_{n-1}(t)D^{\lambda_{n-1}}\overline{y}(t;r) + \cdots + a_1 D^{\lambda_1}\overline{y}(t;r) + a_0(t)D^{\lambda_0}\overline{y}(t;r) = \overline{g}(t;r),$$
(3.4)

subject to fuzzy initial condition

$$\left[\underline{y}^{(k)}(0;r), \overline{y}^{(k)}(0;r)\right] = \left[\underline{b_k}(r), \overline{b}_k(r)\right], \quad k = 0, 1, 2, \dots, n-1,$$

$$n - 1 \leq \lambda < n, \text{ where } r \in [0,1].$$

Case 2   Coefficients $a_{n-1}(t), a_{n-2}(t), \dots, a_1(t), a_0(t)$ are all negative. From Eq. (3.2), we have

$$D^{\lambda_n}\underline{y}(t;r) + a_{n-1}(t)D^{\lambda_{n-1}}\overline{y}(t;r) + \cdots + a_1(t)D^{\lambda_1}\overline{y}(t;r) + a_0(t)D^{\lambda_0}\overline{y}(t;r) = \underline{g}(t;r),$$
(3.5)

$$D^{\lambda_n}\overline{y}(t;r) + a_{n-1}(t)D^{\lambda_{n-1}}\underline{y}(t;r) + \cdots + a_1(t)D^{\lambda_1}\underline{y}(t;r) + a_0(t)D^{\lambda_0}\underline{y}(t;r) = \overline{g}(t;r),$$
(3.6)

subject to fuzzy initial condition

$$\left[\underline{y}^{(k)}(0;r), \overline{y}^{(k)}(0;r)\right] = \left[\underline{b_k}(r), \overline{b}_k(r)\right], \quad k = 0, 1, 2, \dots, n,$$

$$n - 1 \leq \lambda < n, \text{ where } r \in [0,1].$$

Case 3   Coefficients $a_{n-1}(t), \dots, a_{n-m}(t)$ are positive and $a_{n-m-1}(t), a_{n-m-2}(t),$ $\dots, a_1(t), a_0(t)$ are negative From Eq. (3.2), we have

$$D^{\lambda_n}\underline{y}(t;r) + a_{n-1}(t)D^{\lambda_{n-1}}\underline{y}(t;r) + \cdots + a_{n-m}(t)D^{\lambda_{n-m}}\underline{y}(t;r)$$

$$+ a_{n-m-1}(t)D^{\lambda_{n-m-1}}\overline{y}(t;r) + \cdots + a_0(t)D^{\lambda_0}\overline{y}(t;r) = \underline{g}(t;r),$$
(3.7)

and

$$D^{\lambda_n}\overline{y}(t;r) + a_{n-1}(t)D^{\lambda_{n-1}}\overline{y}(t;r) + \cdots + a_{n-m}D^{\lambda_{n-m}}\overline{y}(t;r)$$

$$+ a_{n-m-1}D^{\lambda_{n-m-1}}\underline{y}(t;r) + \cdots + a_0(t)D^{\lambda_0}\underline{y}(t;r) = \overline{g}(t;r),$$
(3.8)

subject to fuzzy initial condition

$$\left[\underline{y}^{(k)}(0;r), \overline{y}^{(k)}(0;r)\right] = \left[\underline{b_k}(r), \overline{b}_k(r)\right], \quad k = 1, 2, \dots, n,$$

$$n - 1 \leq \lambda < n, \text{ where } r \in [0,1].$$

Here, three analytical methods have been proposed to obtain the solution of n-term FFDEs.

3.2 PROPOSED METHODS

Method 1 Fuzzy-Center-Based Method (FCM) (Tapaswini and Chakraverty, 2014a) Here, fuzzy center has been used to solve n-term fuzzy linear fractional differential equations with respect to three cases. First, the fuzzy center solution is obtained and then the lower bound is written in terms of fuzzy center, from which we may find the upper bound of the fuzzy solution. Similarly, the lower bound can be obtained.

Case 1 Coefficients $a_{n-1}(t), a_{n-2}(t), ..., a_1(t), a_0(t)$ are all positive. First, we will write Eq. (3.1) in terms of fuzzy center as

$$D^{\lambda_n}\widetilde{y}^c(t; r) - a_{n-1}(t)D^{\lambda_{n-1}}\widetilde{y}^c(t; r) + \cdots + a_1(t)D^{\lambda_1}\widetilde{y}^c(t; r) + a_0(t)D^{\lambda_0}\widetilde{y}^c(t; r)$$
$$= \widetilde{g}^c(t; r), \tag{3.9}$$

with initial condition

$$(\widetilde{y}^c)^{(k)}(0) = \widetilde{b}_k^c, \quad k = 0, 1, 2, ..., n-1, \quad n-1 \leq \lambda < n.$$

Equation (3.9) may easily be solved to obtain $\widetilde{y}^c(t)$ by applying Laplace transform of Caputo derivative. Now, solving Eq. (3.3) or (3.4), one may get $\underline{y}(t; r)$ or $\overline{y}(t; r)$, respectively. Next, substituting the aforementioned value of $\widetilde{y}^c(t)$ and $\underline{y}(t; r)$ or $\overline{y}(t; r)$ to the definition of fuzzy center, we may find the solution as $\overline{y}(t; r) = 2\widetilde{y}^c(t) - \underline{y}(t; r)$ or $\overline{y}(t; r) = 2\widetilde{y}^c(t) - \underline{y}(t; r)$.

Case 2 Coefficients $a_{n-1}(t), a_{n-2}(t), ..., a_1(t), a_0(t)$ are all negative. Equation (3.1) may be written in terms of fuzzy center as

$$D^{\lambda_n}\widetilde{y}^c(t; r) - a_{n-1}(t)D^{\lambda_{n-1}}\widetilde{y}^c(t; r) - \cdots - a_1(t)D^{\lambda_1}\widetilde{y}^c(t; r) - a_0(t)D^{\lambda_0}\widetilde{y}^c(t; r) = \widetilde{g}^c(t; r), \tag{3.10}$$

with initial condition

$$(\widetilde{y}^c)^{(k)}(0) = \widetilde{b}_k^c, \quad k = 0, 1, 2, ..., n-1, \quad n-1 \leq \lambda < n.$$

$\widetilde{y}^c(t)$ may be obtained by solving Eq. (3.10).

Using the definition of fuzzy center, one may write Eqs. (3.5) and (3.6) as

$$D^{\lambda_n}\underline{y}(t; r) + a_{n-1}(t)D^{\lambda_{n-1}}\left(2\widetilde{y}^c(t; r) - \underline{y}(t; r)\right) + \cdots$$
$$+ a_1(t)D^{\lambda_1}\left(2\widetilde{y}^c(t; r) - \underline{y}(t; r)\right)$$
$$+ a_0(t)D^{\lambda_0}\left(2\widetilde{y}^c(t; r) - \underline{y}(t; r)\right) = \underline{g}(t; r), \tag{3.11}$$

and

$$D^{\lambda_n}\overline{y}(t; r) + a_{n-1}(t)D^{\lambda_{n-1}}\left(2y^c(t; r) - \overline{y}(t; r)\right) + \cdots + a_1(t)D^{\lambda_1}\left(2y^c(t; r) - \overline{y}(t; r)\right)$$
$$+ a_0(t)D^{\lambda_0}\left(2y^c(t; r) - \overline{y}(t; r)\right) = \overline{g}(t; r), \tag{3.12}$$

It may be seen that the aforementioned differential equations are now crisp differential equations. Hence, solving one of the aforementioned crisp differential equations, one may get the solution as $\underline{y}(t; r)$ or $\overline{y}(t; r)$. Applying the definition of fuzzy center, one may get $\overline{y}(t; r) = \left(2\widetilde{y}^{c} - \underline{y}\right)$ or $\underline{y}(t; r) = \left(2\widetilde{y}^{c} - \overline{y}\right)$.

Case 3   Coefficients $a_{n-1}(t), \ldots, a_{n-m}(t)$ are positive and $a_{n-m-1}(t), a_{n-m-2}(t),$ $\ldots, a_1(t), a_0(t)$ are negative. In this case, we may write Eq. (3.1) in terms of fuzzy center as

$$D^{\lambda_n}\widetilde{y}^{c}(t; r) + a_{n-1}(t) D^{\lambda_{n-1}}\widetilde{y}(t; r) + \cdots + a_{n-m}(t) D^{\lambda_{n-m}}\widetilde{y}(t; r)$$

$$- a_{n-m-1}(t) D^{\lambda_{n-m-1}}\widetilde{y}(t; r) - \cdots - a_0(t) D^{\lambda_0}\widetilde{y}(t; r) = \widetilde{g}^{c}(t; r), \qquad (3.13)$$

with initial conditions

$$\left(\widetilde{y}^{c}\right)^{(k)}(0) = \widetilde{b}_{k}^{c}, \quad k = 0, 1, 2, \ldots, n-1, \quad n-1 \leq \lambda < n.$$

Similarly, we may have the solutions for $\widetilde{y}^{c}(t; r)$.

Next, Eqs. (3.7) and (3.8) are written as

$$D^{\lambda_n}\underline{y}(t; r) + a_{n-1}(t) D^{\lambda_{n-1}}\underline{y}(t; r) + \cdots + a_{n-m}(t) D^{\lambda_{n-m}}\underline{y}(t; r) + a_{n-m-1}(t) D^{\lambda_{n-m-1}}$$

$$\left(2y^{c}(t; r) - \underline{y}(t; r)\right) + \cdots + a_0(t) D^{\lambda_0}\left(2y^{c}(t; r) - \underline{y}(t; r)\right) = \underline{g}(t; r), \qquad (3.14)$$

and

$$D^{\lambda_n}\overline{y}(t; r) + a_{n-1}(t) D^{\lambda_{n-1}}\overline{y}(t; r) + \cdots + a_{n-m}D^{\lambda_{n-m}}\overline{y}(t; r) + a_{n-m-1}D^{\lambda_{n-m-1}}$$

$$\left(2y^{c}(t; r) - \overline{y}(t; r)\right) + \cdots + a_0(t) D^{\lambda_0}\left(2y^{c}(t; r) - \overline{y}(t; r)\right) = \overline{g}(t; r). \qquad (3.15)$$

Similarly to the previous cases, one may obtain the lower and upper bounds of the solution by solving Eqs. (3.14) and (3.15), respectively, using the value of fuzzy center $\left(\widetilde{y}^{c}(t; r)\right)$. Otherwise, only one equation, namely Eq. (3.14) or (3.15), may be solved, and using the expression $\overline{y}(t; r) = \left(2\widetilde{y}^{c}(t; r) - \underline{y}(t; r)\right)$ or $\underline{y}(t; r) = \left(2\widetilde{y}^{c}(t; r) - \overline{y}(t; r)\right)$, one may have the solution bounds.

In the following paragraphs, example problems are solved using the proposed Method 1 (FCM) and are also compared with the existing crisp solutions.

Example 3.1   Let us consider the following (Case 1) linear FFDE:

$$D^{\alpha}\widetilde{y}(t; r) = \widetilde{y}(t; r), \quad 0 < \alpha \leq 1, \quad r \in [0, 1] \qquad (3.16)$$

subject to fuzzy initial conditions

$$\widetilde{y}(0) = [0.2r + 0.8, 1.2 - 0.2r].$$

For $r = 1$, the crisp solutions is obtained by the method of Kazem (2013) as

$$y(t) = E_\alpha(t^\alpha).$$

According to Eq. (3.9), the differential equation (Eq. (3.16)) can be written as

$$D^\alpha \tilde{y}^c(t; r) = \tilde{y}^c(t; r). \qquad (3.17)$$

Solving Eq. (3.17), one may obtain $\tilde{y}^c(t; r) = E_\alpha(t^\alpha)$. Proceeding as Eq. (3.3) or (3.4) with the aforementioned value of $\tilde{y}^c(t; r)$ and solving any one of those equations, we obtain the value of $\underline{y}(t; r) = (0.2r + 0.8) E_\alpha(t^\alpha)$ and $\overline{y}(t; r) = (1.2 - 0.2r) E_\alpha(t^\alpha)$.

Hence, one may obtain the final solution as $\tilde{y}(t; r) = \left[\underline{y}(t; r), \overline{y}(t; r)\right]$, where

$$\overline{y}(t; r) = (0.2r + 0.8) E_\alpha(t^\alpha),$$

and

$$\underline{y}(t; r) = (1.2 - 0.2r) E_\alpha(t^\alpha).$$

Example 3.2   We consider the following (Case 2) linear FFDE:

$$D^\alpha \tilde{y}(t; r) = \frac{2}{\Gamma(3 - \alpha)} t^{2-\alpha} - \frac{1}{\Gamma(2 - \alpha)} t^{1-\alpha} - \tilde{y}(t; r) + t^2 - t,$$

$$t > 0, \quad 0 < \alpha \le 1, \quad r \in [0, 1]$$

subject to the fuzzy initial condition

$$\tilde{y}(0) = [0.2r - 0.2, 0.2 - 0.2r].$$

We can find the center solution as $\tilde{y}^c(t; r) = t^2 - t$. Subsequently, by applying Method 1 (FCM), we get the solution

$$\underline{y}(t; r) = (0.2r - 0.2) E_\alpha(t^\alpha) + t^2 - t,$$

and

$$\overline{y}(t; r) = (0.2 - 0.2r) E_\alpha(t^\alpha) + t^2 - t.$$

It is interesting to note that the lower and upper bounds are the same for $r = 1$ of Examples 3.1 and 3.2, which show complete agreement with the crisp solution obtained by Kazem (2013) and Odibat and Momani (2008a). Also, the fuzzy solutions of both the examples exactly match with that of the method of Prakash et al. (2015).

Method 2   Method Based on Addition and Subtraction of Fuzzy Numbers (ASFM) This method is based on addition and subtraction of fuzzy numbers to solve n-term fuzzy fractional linear differential equations, and the methods for each of the three cases are given as follows.

Case 1   Coefficients $a_{n-1}(t), a_{n-2}(t), \ldots, a_1(t), a_0(t)$ are all positive. First, one may write Eq. (3.2) as

$$\left[D^{\lambda_n} \underline{y}(t;r) + D^{\lambda_n} \overline{y}(t;r) \right] + a_{n-1}(t) \left[D^{\lambda_{n-1}} \underline{y}(t;r) + D^{\lambda_{n-1}} \overline{y}(t;r) \right] + \cdots$$

$$+ a_1 \left[D^{\lambda_1} \underline{y}(t;r) + D^{\lambda_1} \overline{y}(t;r) \right] + a_0(t) \left[D^{\lambda_0} \underline{y}(t;r) + D^{\lambda_0} \overline{y}(t;r) \right]$$

$$= \left[\underline{g}(t;r) + \overline{g}(t;r) \right], \qquad (3.18)$$

and

$$\left[D^{\lambda_n} \underline{y}(t;r) - D^{\lambda_n} \overline{y}(t;r) \right] + a_{n-1}(t) \left[D^{\lambda_{n-1}} \underline{y}(t;r) - D^{\lambda_{n-1}} \overline{y}(t;r) \right] + \cdots$$

$$+ a_1 \left[D^{\lambda_1} \underline{y}(t;r) - D^{\lambda_1} \overline{y}(t;r) \right] + a_0(t) \left[D^{\lambda_0} \underline{y}(t;r) - D^{\lambda_0} \overline{y}(t;r) \right]$$

$$= \left[\underline{g}(t;r) - \overline{g}(t;r) \right], \qquad (3.19)$$

with initial condition

$$\left[\underline{y}^{(k)}(0;r) + \overline{y}^{(k)}(0;r) \right] = \left[\underline{b}_k(r) + \overline{b}_k(r) \right], \quad k = 0, 1, 2, \ldots, n-1, \ \ n-1 \leq \lambda < n$$
$$(3.20)$$

and

$$\left[\underline{y}^{(k)}(0;r) - \overline{y}^{(k)}(0;r) \right] = \left[\underline{b}_k(r) - \overline{b}_k(r) \right], \quad k = 1, 2, \ldots, n, \ \ n-1 \leq \lambda < n,$$
$$(3.21)$$

respectively.

Let us now denote

$$\left[D^{\lambda_n} \underline{y}(t;r) + D^{\lambda_n} \overline{y}(t;r) \right] = D^{\lambda_n} \widetilde{u}(t;r),$$

$$\left[D^{\lambda_{n-1}} \underline{y}(t;r) + D^{\lambda_{n-1}} \overline{y}(t;r) \right] = D^{\lambda_{n-1}} \widetilde{u}(t;r),$$

$$\left[D^{\lambda_1} \underline{y}(t;r) + D^{\lambda_1} \overline{y}(t;r) \right] = D^{\lambda_1} \widetilde{u}(t;r), \quad \left[D^{\lambda_0} \underline{y}(t;r) + D^{\lambda_0} \overline{y}(t;r) \right] = D^{\lambda_0} \widetilde{u}(t;r),$$

$$\left[\underline{g}(t;r) + \overline{g}(t;r) \right] = \widetilde{f}(t;r), \quad \left[\underline{y}^{(k)}(0;r) + \overline{y}^{(k)}(0;r) \right] = \widetilde{u}^{(k)}(0;r),$$

$$\left[\underline{b}_k(r) + \overline{b}_k(r) \right] = \widetilde{b}_k(r) \quad \text{for} \quad k = 0, 1, 2, \ldots, n-1, \ \ n-1 \leq \lambda < n,$$

and

$$\left[D^{\lambda_n} \underline{y}(t;r) - D^{\lambda_n} \bar{y}(t;r) \right] = D^{\lambda_n} \widetilde{v}(t;r),$$

$$\left[D^{\lambda_{n-1}} \underline{y}(t;r) + D^{\lambda_{n-1}} \bar{y}(t;r) \right] = D^{\lambda_{n-1}} \widetilde{v}(t;r),$$

$$\left[D^{\lambda_1} \underline{y}(t;r) - D^{\lambda_1} \bar{y}(t;r) \right] = D^{\lambda_1} \widetilde{v}(t;r), \quad \left[D^{\lambda_0} \underline{y}(t;r) - D^{\lambda_0} \bar{y}(t;r) \right] = D^{\lambda_0} \widetilde{v}(t;r),$$

$$\left[\underline{g}(t;r) - \bar{g}(t;r) \right] = \widetilde{g}(t;r), \quad \left[\underline{y}^{(k)}(0;r) - \bar{y}^{(k)}(0;r) \right] = \widetilde{v}^{(k)}(0;r),$$

$$\left[\underline{b}_k(r) - \bar{b}_k(r) \right] = \widetilde{c}_k(r) \quad \text{for} \quad k = 0, 1, 2, \dots, n, \quad n - 1 \leq \lambda < n.$$

Substituting these values in Eqs. (3.18)–(3.21), respectively, we get

$$D^{\lambda_n} \widetilde{u}(t;r) + a_{n-1}(t) D^{\lambda_{n-1}} \widetilde{u}(t;r) + \cdots + a_1(t) D^{\lambda_1} \widetilde{u}(t;r) + a_0(t) D^{\lambda_0} \widetilde{u}(t;r) = \widetilde{f}(t;r), \tag{3.22}$$

and

$$D^{\lambda_n} \widetilde{v}(t;r) + a_{n-1}(t) D^{\lambda_{n-1}} \widetilde{v}(t;r) + \cdots + a_1(t) D^{\lambda_1} \widetilde{v}(t;r) + a_0(t) D^{\lambda_0} \widetilde{v}(t;r) = \widetilde{g}(t;r), \tag{3.23}$$

subject to initial condition

$$\widetilde{u}^{(k)}(0;r) = \widetilde{b}_k(r), \quad \text{for} \quad k = 0, 1, 2, \dots, n - 1, \quad n - 1 \leq \lambda < n,$$

and

$$\widetilde{v}^{(k)}(0;r) = \widetilde{c}_k(r) \quad \text{for} \quad k = 0, 1, 2, \dots, n - 1, \quad n - 1 \leq \lambda < n,$$

respectively.

It may be seen that the aforementioned differential equations are now crisp differential equations. Hence, solving Eqs. (3.22) and (3.23) by any standard method, one may get the solution $\widetilde{u}(t;r)$ and $\widetilde{v}(t;r)$, where $\widetilde{u}(t;r) = \underline{y}(t;r) + \bar{y}(t;r)$ and $\widetilde{v}(t;r) = \underline{y}(t;r) + \bar{y}(t;r)$. Now solving the aforementioned system of equation, one may get the solution of Eq. (3.1).

Case 2 **Coefficients $a_{n-1}(t), a_{n-2}(t), \dots a_1(t), a_0(t)$ are all negative.** One may write Eq. (3.2) as

$$\left[D^{\lambda_n} \underline{y}(t;r) + D^{\lambda_n} \bar{y}(t;r) \right] + |a_{n-1}(t)| \left[D^{\lambda_{n-1}} \underline{y}(t;r) + D^{\lambda_{n-1}} \bar{y}(t;r) \right] + \cdots$$

$$+ |a_1(t)| \left[D^{\lambda_1} \underline{y}(t;r) + D^{\lambda_1} \bar{y}(t;r) \right] + |a_0(t)| \left[D^{\lambda_0} \underline{y}(t;r) + D^{\lambda_0} \bar{y}(t;r) \right]$$

$$= \left[\underline{f}(t;r) + \bar{f}(t;r) \right], \tag{3.24}$$

and

$$\left[D^{\lambda_n}\underline{y}(t;r) - D^{\lambda_n}\overline{y}(t;r)\right] + |a_{n-1}(t)| \left[D^{\lambda_{n-1}}\underline{y}(t;r) - D^{\lambda_{n-1}}\overline{y}(t;r)\right] + \cdots$$

$$+ |a_1(t)| \left[D^{\lambda_1}\underline{y}(t;r) - D^{\lambda_1}\overline{y}(t;r)\right] + |a_0(t)| \left[D^{\lambda_0}\underline{y}(t;r) - D^{\lambda_0}\overline{y}(t;r)\right]$$

$$= \left[\underline{g}(t;r) - \overline{g}(t;r)\right], \tag{3.25}$$

with initial condition

$$\widetilde{u}^{(k)}(0;r) = \widetilde{b}_k(r), \quad \text{for } k = 0, 1, 2, \ldots, n-1, \quad n-1 \le \lambda < n, \tag{3.26}$$

and

$$\widetilde{v}^{(k)}(0;r) = \widetilde{c}_k(r), \quad \text{for } k = 0, 1, 2, \ldots, n-1, \quad n-1 \le \lambda < n, \tag{3.27}$$

respectively.

It may again be seen that the aforementioned differential equations are now crisp differential equations. Hence, solving the aforementioned equations by any known method, one may get the solution as $\widetilde{u}(t;r)$ and $\widetilde{v}(t;r)$, respectively. Then solving the corresponding equations, one may obtain the solutions of the given fuzzy differential equations.

Case 3 Coefficients $a_{n-1}(t), \ldots a_{n-m}(t)$ are positive and $a_{n-m-1}(t), a_{n-m-2}(t),$ $\ldots a_1(t), a_0(t)$ are negative. Similarly to Cases 1 and 2, one may write Eq. (3.2) as

$$\left[D^{\lambda_n}\underline{y}(t;r) + D^{\lambda_n}\overline{y}(t;r)\right] + a_{n-1}(t) \left[D^{\lambda_{n-1}}\underline{y}(t;r) + D^{\lambda_{n-1}}\overline{y}(t;r)\right] + \cdots$$

$$+ a_{n-m}(t) \left[D^{\lambda_{n-m}}\underline{y}(t;r) + D^{\lambda_{n-m}}\overline{y}(t;r)\right]$$

$$+ |a_{n-m-1}(t)| \left[D^{\lambda_{n-m-1}}\underline{y}(t;r) + D^{\lambda_{n-m-1}}\overline{y}(t;r)\right] + \cdots$$

$$+ |a_0(t)| \left[D^{\lambda_0}\underline{y}(t;r) + D^{\lambda_0}\overline{y}(t;r)\right] = \left[\underline{g}(t;r) + \overline{g}(t;r)\right], \tag{3.28}$$

and

$$\left[D^{\lambda_n}\underline{y}(t;r) - D^{\lambda_n}\overline{y}(t;r)\right] + a_{n-1}(t) \left[D^{\lambda_{n-1}}\underline{y}(t;r) - D^{\lambda_{n-1}}\overline{y}(t;r)\right] + \cdots$$

$$+ a_{n-m}(t) \left[D^{\lambda_{n-m}}\underline{y}(t;r) - D^{\lambda_{n-m}}\overline{y}(t;r)\right]$$

$$+ |a_{n-m-1}(t)| \left[D^{\lambda_{n-m-1}}\underline{y}(t;r) - D^{\lambda_{n-m-1}}\overline{y}(t;r)\right] + \cdots$$

$$+ |a_0(t)| \left[D^{\lambda_0}\underline{y}(t;r) - D^{\lambda_0}\overline{y}(t;r)\right] = \left[\underline{g}(t;r) - \overline{g}(t;r)\right], \tag{3.29}$$

with initial condition

$$\widetilde{u}^{(k)}(0;r) = \widetilde{b}_k(r), \quad \text{for} \quad k = 0, 1, 2, \dots, n-1, \quad n-1 \le \lambda < n, \quad (3.30)$$

and

$$\widetilde{v}^{(k)}(0;r) = \widetilde{c}_k(r) \quad \text{for} \quad k = 1, 2, \dots, n, \quad n-1 \le \lambda < n, \quad (3.31)$$

respectively.

By applying the procedure as discussed in Cases 1 and 2, we get the solution of the given fuzzy fractional differential equations.

In the following paragraphs, example problems are solved using the proposed Method 2 (ASFM) and are also compared with crisp solutions obtained by the method of Kazem (2013).

Example 3.3   Consider Example 3.1 (Case 1). Subject to the Gaussian fuzzy initial conditions,

$$\widetilde{y}(0;r) = \left[1 - 0.1\sqrt{-2\log r}, 1 + 0.1\sqrt{-2\log r}\right].$$

The crisp solution is obtained by

$$y(t) = E_\alpha(t^\alpha).$$

Now using the proposed method (ASFM) (Case 1) proceeding as Eqs. (3.22) and (3.23), we have

$$\widetilde{u}(t;r) = \underline{y}(t;r) + \overline{y}(t;r) = 2E_\alpha(t^\alpha), \quad (3.32)$$

$$\widetilde{v}(t;r) = \underline{y}(t;r) - \overline{y}(t;r) = -0.2\sqrt{-2\log r}E_\alpha(t^\alpha). \quad (3.33)$$

Solving the corresponding Eqs. (3.32) and (3.33), one may get

$$\underline{y}(t;r) = \left(1 - 0.1\sqrt{-2\log r}\right)E_\alpha(t^\alpha),$$

$$\overline{y}(t;r) = \left(1 + 0.1\sqrt{-2\log r}\right)E_\alpha(t^\alpha).$$

Example 3.4   Let us consider now the following linear FFDE:

$$D^\alpha \widetilde{y}(t;r) + \widetilde{y}(t;r) = 0, \quad 0 < \alpha \le 1, \ r \in [0, 1] \quad (3.34)$$

subject to the triangular fuzzy initial conditions as

$$\widetilde{y}(0;r) = [0.2r + 0.8, 1.2 - 0.2r]$$

and

$$\overline{y}'\,(0;r) = [0.1r - 0.1, 0.1 - 0.1r].$$

Again using the proposed method, we have

$$\widetilde{u}\,(t;r) = \underline{y}\,(t;r) + \overline{y}\,(t;r) = 2E_\alpha\,(-t^\alpha),\qquad(3.35)$$

and

$$\widetilde{v}\,(t;r) = \underline{y}\,(t;r) - \overline{y}\,(t;r) = (0.4r - 0.4)\,E_\alpha\,(-t^\alpha) + (0.2r - 0.2)\,tE_{\alpha,2}\,(-t^\alpha).\quad(3.36)$$

Subsequently, solving Eqs. (3.35) and (3.36), we get the solution

$$\underline{y}\,(t;\ r) = (0.2r + 0.8)\,E_\alpha\,(-t^\alpha) + (0.1r - 0.1)\,tE_{\alpha,2}\,(-t^\alpha),$$

$$\overline{y}\,(t;r) = (1.2 - 0.2r)\,E_\alpha\,(-t^\alpha) + (0.1 - 0.1r)\,tE_{\alpha,2}\,(-t^\alpha).$$

From the results of both the examples, it is interesting to see that for $r = 1$ with triangular fuzzy initial condition, the fuzzy results are exactly the same as the solution obtained by Kazem (2013) with crisp initial condition.

Method 3 Double-Parametric-Based Method (DPM) (Tapaswini and Chakraverty, 2013, 2014a). In this method, double parametric form of fuzzy numbers has been used to solve n-term FFDE. Using the single parametric form, namely r-cut form of fuzzy numbers, the n-term FFDE is converted first to an interval-based FFDE. Next, this differential equation is transformed to crisp form by applying double parametric form of fuzzy numbers. Finally, the same is solved by known methods.

Using the double parametric form as discussed in Definition 2.6, Eq. (3.2) system can be represented as

$$\left\{\beta\left(D^{\lambda_n}\overline{y}\,(t;r) - D^{\lambda_n}\underline{y}\,(t;r)\right) + D^{\lambda_n}\underline{y}\,(t;r)\right\}$$

$$+\,a_{n-1}\,(t)\left\{\beta\left(D^{\lambda_{n-1}}\overline{y}\,(t;r) - D^{\lambda_{n-1}}\underline{y}\,(t;r)\right) + D^{\lambda_{n-1}}\underline{y}\,(t;r)\right\} + \cdots$$

$$+\,a_{n-m}\,(t)\left\{\beta\left(D^{\lambda_{n-m}}\overline{y}\,(t;r) - D^{\lambda_{n-m}}\underline{y}\,(t;r)\right) + D^{\lambda_{n-m}}\underline{y}\,(t;r)\right\}$$

$$+\,a_{n-m-1}\,(t)\left\{\beta\left(D^{\lambda_{n-m-1}}\overline{y}\,(t;r) - D^{\lambda_{n-m-1}}\underline{y}\,(t;r)\right) + D^{\lambda_{n-m-1}}\underline{y}\,(t;r)\right\} + \cdots$$

$$+\,a_0\,(t)\left\{\beta\left(D^{\lambda_0}\overline{y}\,(t;r) - D^{\lambda_0}\underline{y}\,(t;r)\right) + D^{\lambda_0}\underline{y}\,(t;r)\right\}$$

$$=\left\{\beta\left(\overline{g}\,(t;\ r) - \underline{g}\,(t;r)\right) + \underline{g}\,(t;r)\right\},\qquad(3.37)$$

subject to the fuzzy initial conditions

$$\left\{\beta\left(\overline{y}(0;r) - \underline{y}(0;r)\right) + \underline{y}(0;r)\right\} = \left\{\beta\left(\overline{b}_0(r) - \underline{b}_0(r)\right) + \underline{b}_0(r)\right\},$$

$$\left\{\beta\left(\left(\overline{y}'(0;r) - \underline{y}'(0;r)\right)\right) + \underline{y}'(0;r)\right\} = \left\{\beta\left(\overline{b}_1(r) - \underline{b}_1(r)\right) + \underline{b}_1(r)\right\}, \dots,$$

$$\left\{\beta\left(\overline{y}^{(n-1)}(0;r) - \underline{y}^{(n-1)}(0;r)\right) + \underline{y}^{(n-1)}(0;r)\right\}$$

$$= \left\{\beta\left(\overline{b}_{n-1}(r) - \underline{b}_{n-1}(r)\right) + \underline{b}_{n-1}(r)\right\}.$$

Let us now consider

$$\left\{\beta\left(D^{\lambda_j}\overline{y}(t;r) - D^{\lambda_j}\underline{y}(t;r)\right) + D^{\lambda_j}\underline{y}(t;r)\right\} = D^{\lambda_j}\widetilde{y}(t;r,\beta),$$

$$\left\{\beta\left(\overline{g}(t;r) - \underline{g}(t;r)\right) + \underline{g}(t;r)\right\} = \widetilde{g}(t;r,\beta),$$

$$\left\{\beta\left(\overline{y}^{(j)}(0;r) - \underline{y}^{(j)}(0;r)\right) + \underline{y}^{(j)}(0;r)\right\} = \widetilde{y}^{(j)}(0;r,\beta) \quad \text{and}$$

$$\left\{\beta\left(\overline{b}_j(r) - \underline{b}_j(r)\right) + \underline{b}_j(r)\right\} = \widetilde{b}_j(r,\beta) \quad \text{for} \quad j = 0, 1, 2, \dots, n-1.$$

Again substituting these values in Eq. (3.37), one may have

$$D^{\lambda_n}\widetilde{y}(t;r,\beta) + a_{n-1}(t)\widetilde{y}(t;r,\beta) + \cdots + a_{n-m}(t)D^{\lambda_{n-m}}\widetilde{y}(t;r,\beta)$$

$$+ a_{n-m-1}(t)D^{\lambda_{n-m-1}}\widetilde{y}(t;r,\beta) + a_0(t)D^{\lambda_0}\widetilde{y}(t;r,\beta) = \widetilde{g}(t;r,\beta),$$

with fuzzy initial conditions

$$\widetilde{y}(0;r,\beta) = \widetilde{b}_0(r,\beta), \widetilde{y}'(0;r,\beta) = \widetilde{b}_1(r,\beta), \dots, \widetilde{y}^{(n-1)}(0;r,\beta) = \widetilde{b}_{n-1}(r,\beta).$$

Hence, solving the corresponding crisp differential equation, one may get the solution as $\widetilde{y}(t;r,\beta)$. To obtain the solution bound in single parametric form, we may use $\beta = 0$ and 1 to get the lower and the upper bound of the solution, respectively. This may be represented as

$$\widetilde{y}(t;r,0) = \underline{y}(t,r) \quad \text{and} \quad \widetilde{y}(t,r,1) = \overline{y}(t,r).$$

Example 3.5   Consider the following linear fuzzy fractional initial value problem:

$$D^{\alpha}\widetilde{y}(t;r) = \widetilde{y}(t;r) + 1, \quad 0 < \alpha \le 1, \quad r \in [0, 1] \tag{3.38}$$

subject to triangular fuzzy initial condition as

$$\widetilde{y}'(0;r) = [0.2r - 0.2, 0.2 - 0.2r].$$

Now, by following the proposed Method 3 (DPFM), the original differential equation is represented by using a double parametric form of fuzzy numbers as

$$\left\{ \beta \left(D^{\alpha}\overline{y}\,(t; r, \beta) - D^{\alpha}\underline{y}\,(t; r, \beta) \right) + D^{\alpha}\underline{y}\,(t; r, \beta) \right\}$$
$$= \left\{ \beta \left(\overline{y}\,(t; r, \beta) - \underline{y}\,(t; r, \beta) \right) + \underline{y}\,(t; r, \beta) \right\} + 1, \quad 0 < \alpha \le 1, \quad r, \beta \in [0, 1].$$
$$(3.39)$$

subject to the initial conditions

$$\widetilde{y}\,'(0; r, \beta) = \{ \beta\,(0.4 - 0.4r) + (0.2r - 0.2) \}\,.$$

Let us consider

$$\left\{ \beta \left(D^{\alpha}\overline{y}\,(t; r, \beta) - D^{\alpha}\underline{y}\,(t; r, \beta) \right) + D^{\alpha}\underline{y}\,(t; r, \beta) \right\} = D^{\alpha}\widetilde{y}\,(t; r, \beta),$$

$$\left\{ \beta \left(\overline{y}\,(t; r, \beta) - \underline{y}\,(t; r, \beta) \right) + \underline{y}\,(t; r, \beta) \right\} = \widetilde{y}\,(t; r, \beta).$$

Substituting these expressions in Eq. (3.39), one may get

$$D^{\alpha}\widetilde{y}\,(t; r, \beta) = \widetilde{y}\,(t; r, \beta) + 1. \qquad (3.40)$$

Equation (3.40) is a crisp differential equation, which may easily be solved by known method. As such, the solution may be written as

$$\widetilde{y}\,(t; r, \beta) = \{ \beta\,(0.4 - 0.4r) + (0.2r - 0.2) \}\,E_{\alpha}\,(t^{\alpha}) + t^{\alpha}E_{\alpha,\alpha+1}\,(t^{\alpha})\,.$$

Substituting $\beta = 0$ and 1 in $\widetilde{y}\,(t; r, \beta)$, we get the lower and upper bounds of the fuzzy solutions in single parametric form, respectively, as

$$\underline{y}\,(t;\ r, 0) = (0.2r - 0.2)\,E_{\alpha}\,(t^{\alpha}) + t^{\alpha}E_{\alpha,\alpha+1}\,(t^{\alpha})\,,$$

$$\overline{y}\,(t; r, 1) = (0.2 - 0.2r)\,E_{\alpha}\,(t^{\alpha}) + t^{\alpha}E_{\alpha,\alpha+1}\,(t^{\alpha})\,.$$

Example 3.6 Finally, we consider the following linear FFDE:

$$D^{\alpha}\widetilde{y}\,(t; r) + \widetilde{y}\,(t; r) = 0, \quad 0 < \alpha \le 1, \quad r \in [0, 1]$$

subject to the triangular fuzzy initial conditions as

$$\widetilde{y}\,(0; r) = [0.2r + 0.8, 1.2 - 0.2r]\,,$$

and

$$\widetilde{y}\,'(0; r) = [0.1r - 0.1, 0.1 - 0.1r]\,.$$

Applying the proposed Method 3, one may have

$$\widetilde{y}(t; r, \beta) = \{\beta (0.4 - 0.4r) + (0.2r + 0.8)\} E_\alpha (-t^\alpha)$$
$$- \{\beta (0.2 - 0.2r) + (0.1r - 0.1)\} tE_{\alpha,2} (-t^\alpha).$$

Substituting $\beta = 0$ and 1 in $\widetilde{y}(t; r, \beta)$, we get the lower and upper bounds of the fuzzy solutions, respectively, as

$$\widetilde{y}(t; r, 0) = (0.2r + 0.8) E_\alpha (-t^\alpha) - (0.1r - 0.1) tE_{\alpha,2} (-t^\alpha),$$

and

$$\widetilde{y}(t; r, 1) = (1.2 - 0.2r) E_\alpha (-t^\alpha) - (0.1 - 0.1r) tE_{\alpha,2} (-t^\alpha).$$

It is interesting to note that for both the aforementioned examples, the lower and upper bounds of the fuzzy solutions are the same for $r = 1$, which again approximately match with the crisp solution of Kazem (2013).

In this chapter, three analytical methods have been discussed to handle n-term FFDE considering the possible cases. It is interesting to note that in Methods 1 and 2, we do not have to solve the coupled system of differential equations; rather, we do solve the two uncoupled crisp differential equations. Method 3 (DPM) is simple and straightforward as compared to Methods 1 and 2, as it directly solves the original fuzzy differential equations using the developed double parametric form.

BIBLIOGRAPHY

Arshad S, Lupulescu V. Fractional differential equation with the fuzzy initial condition. *Electro J Differ Eq* 2011;**2011**:1–8.

Bede B. Note on "Numerical solutions of fuzzy differential equations by predictor-corrector method". *Inform Sci* 2008;**178**:1917–1922.

Hanss M. *Applied Fuzzy Arithmetic: An Introduction with Engineering Applications*. Berlin: Springer-Verlag; 2005.

Hashim I, Abdulaziz O, Momani S. Homotopy analysis method for fractional IVPs. *Commun Nonlinear Sci Numer Simul* 2009;**14**:674–684.

Kazem S. Exact solution of some linear fractional differential equations by Laplace transform. *Int J Nonlinear Sci* 2013;**16**:3–11.

Miller KS, Ross B. *An Introduction to the Fractional Calculus and Fractional Differential Equations*. New York: John Wiley & Sons; 1993.

Moore RE. *Interval Analysis*. Englewood Cliffs: Prentice Hall; 1966.

Mohammed OH, Fadhel FS, Abdul-Khaleq FA. Differential transform method for solving fuzzy fractional initial value problems. *J Basrah Res (Sci)* 2011;**37**:158–170.

Odibat ZM, Momani S. An algorithm for the numerical solution of differential equations of fractional order. *J Appl Math Inform* 2008a;**26**:15–27.

Odibat Z, Momani S. Analytical comparison between the homotopy perturbation method and variational iteration method for differential equations of fractional order. *Int J Modern Phys* 2008b;**B22**:4041–4058.

Petras I. *Fractional-Order Nonlinear Systems Modeling, Analysis and Simulation.* London, New York: Higher Education Press, Springer Heidelberg Dordrecht; 2011.

Podlubny I. *Fractional Differential Equations.* New York, NY: Academic Press; 1999.

Prakash P, Nieto JJ, Senthilvelavan S, Sudha Priya G. Fuzzy fractional initial value problem. *J Intell Fuzzy Syst* 2015,**28**:2691–2704.

Ross TJ. *Fuzzy Logic with Engineering Applications.* New York: John Wiley & Sons; 2004.

Saadatmandi A, Dehghan M. A new operational matrix for solving fractional order differential equations. *Comput Math Appl* 2010;**59**:1326–1336.

Salah A, Khan M, Gondal MA. A novel solution procedure for fuzzy fractional heat equations by homotopy analysis transform method. *Neural Comput Appl* 2013;**23**:269–271.

Tapaswini S, Chakraverty S. Numerical solution of uncertain beam equations using double parametric form of fuzzy numbers. *Appl Comput Intell Soft Comput* 2013;**2013**:1–8.

Tapaswini S, Chakraverty S. New analytical method for solving *n*-th order fuzzy differential equations. *Ann Fuzzy Math Inform* 2014a;**8**:231–244.

Tapaswini S, Chakraverty S. Dynamic response of imprecisely defined beam subject to various loads using Adomian decomposition method. *Appl Soft Comput* 2014b;**24**:249–263.

Tapaswini S. Numerical solution of fuzzy differential equations [PhD Thesis]. India: National Institute of Technology Rourkela; 2014.

Zadeh LA. Fuzzy sets. *Inform Control* 1965;**8**:338–353.

Zimmermann HJ. *Fuzzy Set Theory and Its Application.* Boston/Dordrecht/London: Kluwer Academic Publishers; 2001.

4

NUMERICAL METHODS FOR FUZZY FRACTIONAL DIFFERENTIAL EQUATIONS

In general, it is not always possible to obtain the exact solutions of Fuzzy Fractional Differential Equations (FFDEs). So, numerical methods are used to obtain the solution of FFDEs. This chapter investigates numerical schemes, namely homotopy perturbation method (HPM), Adomian decomposition method (ADM), and variational iteration method (VIM), and these are implemented to solve fuzzy ODEs and PDEs.

4.1 HOMOTOPY PERTURBATION METHOD (HPM) (HE, 1999a, 2000a)

To illustrate the basic idea of this method, we consider the following nonlinear differential equation of the form

$$A(u) - f(a) = 0, \quad a \in \Omega, \tag{4.1}$$

with the boundary condition

$$B\left(u, \frac{\partial u}{\partial n}\right) = 0, \quad t \in \Gamma, \tag{4.2}$$

where A is a general differential operator, B a boundary operator, $f(a)$ a known analytical function, and Γ is the boundary of the domain Ω. A can be divided into two

Fuzzy Arbitrary Order System: Fuzzy Fractional Differential Equations and Applications, First Edition.
Snehashish Chakraverty, Smita Tapaswini, and Diptiranjan Behera.
© 2016 John Wiley & Sons, Inc. Published 2016 by John Wiley & Sons, Inc.

parts, which are L and N, where L is linear and N is nonlinear. Therefore, Eq. (4.1) may be written as follows:

$$L(u) + N(u) - f(a) = 0, \quad a \in \Omega. \tag{4.3}$$

By the homotopy technique, we construct a homotopy $U(a,p) : \quad \Omega \times [0, 1] \to R$, which satisfies the following:

$$H(U,p) = (1-p)\left[L(U) - L(v_0)\right] + p\left[A(U) - f(a)\right] = 0, \ p \in [0,1], \quad r \in \Omega, \tag{4.4}$$

or

$$H(U,p) = L(U) - L(v_0) + pL(v_0) + p\left[N(U) - f(a)\right] = 0, \tag{4.5}$$

where $a \in \Omega$ and $p \in [0, 1]$ is an embedding parameter, v_0 is an initial approximation of Eq. (4.1). Hence, it is obvious that

$$H(U,0) = L(U) - L(v_0) = 0, \tag{4.6}$$

$$H(U,1) = A(U) - f(a) = 0, \tag{4.7}$$

and the changing process of p from 0 to 1 is just that of $U(a,p)$ from $v_0(a)$ to $u(a)$. In topology, this is called deformation, and $L(U) - L(v_0)$, $A(U) - f(a)$ are called homotopic. If the embedding parameter p; $(0 \le p \le 1)$ is considered as a small parameter, applying the classical perturbation technique, we can naturally assume that the solution of Eq. (4.4) or (4.5) can be given as a power series in p as follows:

$$U = u_0 + pu_1 + p^2u_2 + p^3u_3 + \cdots \tag{4.8}$$

when $p \to 1$, Eq. (4.4) or (4.5) corresponds to Eqs. (4.3). Equation (4.8) becomes the approximate solution of Eq. (4.3), that is,

$$u = \lim_{p \to 1} U = u_0 + u_1 + u_2 + u_3 + \cdots. \tag{4.9}$$

The convergence of the series (4.9) has been proved in He (1999a, 2000a).

We now apply the HPM for solving linear fuzzy fractional differential equations.

Example 4.1   Let us consider the linear FFDEs as in Kazem (2013)

$$D^\alpha \widetilde{y}(t) + \widetilde{y}(t) = 0, \quad 0 < \alpha \le 1, \tag{4.10}$$

subject to the triangular fuzzy initial conditions

$$\widetilde{y}(0) = (0.8, 1, 1.2) \text{ and } \widetilde{y}'(0) = (-0.2, 0, 0.2).$$

As per the interval form we may write Eq. (4.10) as

$$\left[D^{\alpha} \underline{y}(t;r), \ D^{\alpha} \overline{y}(t;r) \right] + \left[\underline{y}(t;r), \overline{y}(t;r) \right] = 0, \quad 0 < \alpha \le 1, \ r \in [0,1] \quad (4.11)$$

or

$$D^{\alpha} \widetilde{y}(t;r) + \widetilde{y}(t;r) = 0, \quad 0 < \alpha \le 1, \ r \in [0,1] \quad (4.12)$$

subject to fuzzy initial condition

$$\widetilde{y}(0;r) = [0.2r + 0.8, 1.2 - 0.2r] \quad (4.13)$$

and

$$\widetilde{y}'(0;r) = [0.2r - 0.2, 0.2 - 0.2r]. \quad (4.14)$$

Also, we may write the fuzzy initial condition as follows:

$$\underline{y}(0;r) = 0.2r + 0.8, \ \overline{y}(0;r) = 1.2 - 0.2r \ \text{ and}$$

$$\underline{y}'(0;r) = 0.2r - 0.2, \ \overline{y}'(0;r) = 0.2 - 0.2r.$$

According to HPM, we may construct a simple homotopy for an embedding parameter $p \in [0,1]$ as follows:

$$(1 - p) D^{\alpha} \widetilde{y}(t) + p \left(D^{\alpha} \widetilde{y}(t) + \widetilde{y}(t) \right) = 0, \quad (4.15)$$

or

$$D^{\alpha} \widetilde{y}(t) + p \widetilde{y}(t) = 0, \quad (4.16)$$

In the changing process from 0 to 1, namely for $p = 0$, Eq. (4.15) or (4.16) gives $D^{\alpha} \widetilde{y}(t) = 0$, whereas for $p = 1$, we have the original system

$$D^{\alpha} \widetilde{y}(t;r) + \widetilde{y}(t;r) = 0.$$

This is called deformation in topology. Moreover, $D^{\alpha} \widetilde{y}(t;r)$ and $D^{\alpha} \widetilde{y}(t;r) + \widetilde{y}(t;r)$ are called homotopic. Next, we can assume the solution of Eq. (4.15) or (4.16) as a power series expansion in p as

$$\widetilde{y}(t;r) = \widetilde{y}_0(t;r) + p \widetilde{y}_1(t;r) + p^2 \widetilde{y}_2(t;r) + p^3 \widetilde{y}_3(t;r) + \cdots, \quad (4.17)$$

where, $\widetilde{y}_i(t;r)$ for $i = 0, 1, 2, 3, \ldots$ are functions yet to be determined. Substituting Eq. (4.17) into Eq. (4.15) or (4.16) and equating the terms with the identical powers of p, we have

$$p^0 : \quad D^{\alpha} \widetilde{y}_0(t;r) = 0, \quad (4.18)$$

$$p^1 : \quad D^{\alpha} \widetilde{y}_1(t;r) + \widetilde{y}_0(t;r) = 0, \quad (4.19)$$

$$p^2 : \quad D^\alpha \tilde{y}_2 (t; r) + \tilde{y}_1 (t; r) = 0, \tag{4.20}$$

$$p^3 : \quad D^\alpha \tilde{y}_3 (t; r) + \tilde{y}_2 (t; r) = 0, \tag{4.21}$$

and so on.

Choosing initial approximation $\tilde{y}(0; r)$ and applying the operator J^α (the inverse operator of Caputo derivative D^α) on both sides of Eqs. (4.18)–(4.21), one may obtain the following equations:

$$\tilde{y}_0 (t; r) = \left[\underline{y}_0 (t; r), \bar{y}_0 (t; r) \right] = [0.2r + 0.8, 1.2 - 0.2r],$$

$$\tilde{y}_1 (t; r) = \left[\underline{y}_1 (t; r), \bar{y}_1 (t; r) \right] = \left[-(0.2r + 0.8) \frac{t^\alpha}{\Gamma(\alpha + 1)}, -(1.2 - 0.2r) \frac{t^\alpha}{\Gamma(\alpha + 1)} \right],$$

$$\tilde{y}_2 (t; r) = \left[\underline{y}_2 (t; r), \bar{y}_2 (t; r) \right] = \left[(0.2r + 0.8) \frac{t^{2\alpha}}{\Gamma(2\alpha + 1)}, (1.2 - 0.2r) \frac{t^{2\alpha}}{\Gamma(2\alpha + 1)} \right],$$

$$\tilde{y}_3 (t; r) = \left[\underline{y}_3 (t; r), \bar{y}_3 (t; r) \right] = \left[-(0.2r + 0.8) \frac{t^{3\alpha}}{\Gamma(3\alpha + 1)}, -(1.2 - 0.2r) \frac{t^{3\alpha}}{\Gamma(3\alpha + 1)} \right],$$

and so on.

Now substituting these terms in Eq. (4.17) with $p \to 1$, we get the approximate solution of Eq. (4.12) as

$$\tilde{y}(t; r) = \left[\underline{y}(t; r), \bar{y}(t; r) \right] = \tilde{y}_0 (t; r) + \tilde{y}_1 (t; r) + \tilde{y}_2 (t; r) + \tilde{y}_3 (t; r) + \cdots . \tag{4.22}$$

Therefore, one may have the bound solution as

$$\underline{y}(t; r) = (0.2r + 0.8) \left(1 - \frac{t^\alpha}{\Gamma(\alpha + 1)} + \frac{t^{2\alpha}}{\Gamma(2\alpha + 1)} - \frac{t^{3\alpha}}{\Gamma(3\alpha + 1)} + \cdots \right)$$

$$= (0.2r + 0.8) E_\alpha (-t^\alpha),$$

and

$$\bar{y}(t; r) = (1.2 - 0.2r) \left(1 - \frac{t^\alpha}{\Gamma(\alpha + 1)} + \frac{t^{2\alpha}}{\Gamma(2\alpha + 1)} - \frac{t^{3\alpha}}{\Gamma(3\alpha + 1)} + \cdots \right)$$

$$= (1.2 - 0.2r) E_\alpha (-t^\alpha).$$

One may note that in the special case when $r = 1$, the results (crisp) obtained by the proposed method are exactly the same as that of the solution obtained by Kazem (2013).

4.2   ADOMIAN DECOMPOSITION METHOD (ADM) (ADOMIAN, 1984, 1994)

To illustrate the basic idea of this method, we consider the following differential equation of the form:

$$Fu = g, \tag{4.23}$$

where, F is a general differential operator. F can be divided into three parts, which are L, R, and N, where L is an easily or trivially invertible linear operator, R is the remaining linear part, and N represents a nonlinear operator. Equation (4.23) can therefore be written as follows:

$$Lu + Ru + Nu = g, \tag{4.24}$$

We may write Eq. (4.24) as

$$Lu = g - Ru - Nu. \tag{4.25}$$

Applying the inverse operator L^{-1} (which is the inverse of the operator L) on both sides of Eq. (4.25), one may obtain the following equation:

$$L^{-1}Lu = L^{-1}g - L^{-1}Ru - L^{-1}Nu. \tag{4.26}$$

For initial value problems, we define L^{-1} for $L \equiv \partial^n / \partial t^n$ as the n-fold definite integration operator from 0 to t. For example, if $L \equiv \partial^2 / \partial t^2$ is a second-order operator, then L^{-1} is a twofold integration operator.

We have $L^{-1}Lu = u(t) - u(0) - tu'(0)$, and therefore, Eq. (4.26) becomes

$$u(t) = u(0) + tu'(0) + L^{-1}g - L^{-1}Ru - L^{-1}Nu. \tag{4.27}$$

According to ADM, the solution u is assumed as infinite sum of series (Adomian, 1984, 1994)

$$u = \sum_{n=0}^{\infty} u_n, \tag{4.28}$$

and the nonlinear term Nu is decomposed as follows:

$$Nu = \sum_{n=0}^{\infty} A_n \left(u_0, u_1, u_2, \dots, u_n \right), \tag{4.29}$$

where A_n are the set of Adomian polynomials. A_n's can be found from the formula

$$A_n = \frac{1}{n!} \frac{\partial^n}{\partial \lambda^n} N \left(\sum_{w=1}^{n} \lambda^i u_i \right) \Bigg|_{\lambda=0}. \tag{4.30}$$

Substituting Eqs. (4.28) and (4.29) into Eq. (4.27), we get

$$u = \sum_{n=0}^{\infty} u_n = u_0 - L^{-1}R\sum_{n=0}^{\infty} u_n - L^{-1}\sum_{n=0}^{\infty} A_n, \qquad (4.31)$$

where $u_0 = u(0) + tu'(0) + L^{-1}g$

$$u_{n+1} = -L^{-1}Ru_n - L^{-1}\sum_{n=0}^{\infty} A_n, \quad n \geq 0.$$

Substituting the values of $u_0, u_1, u_2, \ldots$ in Eq. (4.27), one can obtain u. The convergence for the aforementioned series has been found in Cherruault (1989), Abbaoui and Cherruault (1994, 1995), and Himoun et al. (1999).

Now, we apply ADM to solve Example 4.1 with the same fuzzy initial conditions. One may write Eq. (4.12) as

$$D\widetilde{y}(t;r) + D^{1-\alpha}\widetilde{y}(t;r) = 0, \quad 0 < \alpha \leq 1, \ r \in [0,1] \qquad (4.32)$$

We have applied ADM to solve Eq. (4.32) and consider Eq. (4.32) as

$$D\widetilde{y}(t;r) = -D^{1-\alpha}\widetilde{y}(t;r), \quad 0 < \alpha \leq 1 \qquad (4.33)$$

Applying the operator D^{-1} (which is the inverse operator of D) on both sides of Eq. (4.33), the equivalent expression is

$$D^{-1}D\widetilde{y}(t;r) = D^{-1}\left(-D^{1-\alpha}\widetilde{y}(t;r)\right), \quad 0 < \alpha \leq 1 \qquad (4.34)$$

Now, we have
$$D^{-1}D\widetilde{y}(t;r) = \widetilde{y}(t;r) - \widetilde{y}(0;r)$$

Then Eq. (4.34) becomes

$$\widetilde{y}(t;r) = \widetilde{y}(0;r) + D^{-1}\left(-D^{1-\alpha}\widetilde{y}(t;r)\right), \ 0 < \alpha \leq 1 \qquad (4.35)$$

According to Adomian decomposition (Adomian, 1984, 1994), we assume an infinite series solution for unknown function $\widetilde{y}(t;r)$ as

$$\widetilde{y}(t;r) = \sum_{n=0}^{\infty} \widetilde{y}_n(t;r) \qquad (4.36)$$

with $\widetilde{y}_0(t;r) = \widetilde{y}(0;r) = \left[\underline{y}_0(t;r), \bar{y}_0(t;r)\right] = [0.2r + 0.8, 1.2 - 0.2r]$, and the components $\widetilde{y}_n(t;r)$ where $n > 0$ are usually determined by

$$\widetilde{y}_1(t;r) = D^{-1}\left(-D^{1-\alpha}\widetilde{y}_0(t;r)\right),$$

$$\left[\underline{y}_1(t;r), \bar{y}_1(t;r) \right] = \left[-(0.2r+0.8) \frac{t^\alpha}{\Gamma(\alpha+1)}, -(1.2-0.2r) \frac{t^\alpha}{\Gamma(\alpha+1)} \right],$$

$$\tilde{y}_2(t;r) = D^{-1} \left(-D^{1-\alpha} \tilde{y}_1(t;r) \right),$$

$$\left[\underline{y}_2(t;r), \bar{y}_2(t;r) \right] = \left[(0.2r+0.8) \frac{t^{2\alpha}}{\Gamma(2\alpha+1)}, (1.2-0.2r) \frac{t^{2\alpha}}{\Gamma(2\alpha+1)} \right],$$

$$\tilde{y}_3(t;r) = D^{-1} \left(-D^{1-\alpha} \tilde{y}_2(t;r) \right),$$

$$\left[\underline{y}_3(t;r), \bar{y}_3(t;r) \right] = \left[-(0.2r+0.8) \frac{t^{3\alpha}}{\Gamma(3\alpha+1)}, -(1.2-0.2r) \frac{t^{3\alpha}}{\Gamma(3\alpha+1)} \right],$$

and so on.

Now substituting these terms in Eq. (4.36) one may get the approximate solution of Eq. (4.33) as follows:

$$\tilde{y}(t;r) = \tilde{y}_0(t;r) + \tilde{y}_1(t;r) + \tilde{y}_2(t;r) + \tilde{y}_3(t;r) + \cdots . \tag{4.37}$$

Therefore, the solution can be written in general form as

$$y(t;r) = (0.2r+0.8) \left(1 - \frac{t^\alpha}{\Gamma(\alpha+1)} + \frac{t^{2\alpha}}{\Gamma(2\alpha+1)} - \frac{t^{3\alpha}}{\Gamma(3\alpha+1)} + \cdots \right)$$

$$= (0.2r+0.8) E_\alpha (-t^\alpha),$$

and

$$\bar{y}(t;r) = (1.2-0.2r) \left(1 - \frac{t^\alpha}{\Gamma(\alpha+1)} + \frac{t^{2\alpha}}{\Gamma(2\alpha+1)} - \frac{t^{3\alpha}}{\Gamma(3\alpha+1)} + \cdots \right)$$

$$= (1.2-0.2r) E_\alpha (-t^\alpha).$$

4.3   VARIATIONAL ITERATION METHOD (VIM) (HE, 1999b, 2000b)

Again, to illustrate the basic idea of this method, we consider the following general nonlinear system

$$L[u(t)] + N[u(t)] = g(t), \tag{4.38}$$

where L is a linear operator, N is a nonlinear operator, and $g(t)$ is a given continuous function.

The basic character of the method is to construct a correction functional for Eq. (4.38) as follows:

$$u_{n+1}(t) = u_n(t) + \int_0^t \lambda(\tau) \left\{ Lu_n(\tau) + N\tilde{u}_n(\tau) - g(\tau) \right\} d\tau, \tag{4.39}$$

where λ is a general Lagrangian multiplier, which can be identified via variational theory. u_n is the nth approximate solution, and $\tilde{u}_n$ denotes a restricted variation, that is, $\delta \widetilde{u}_n = 0$. The initial approximation u_0 can be freely chosen if it satisfies the initial and boundary conditions of the problem. However, the success of the method depends on the proper selection of the initial approximation u_0. We approximate the solution

$$u(x, t) = \lim_{n \to \infty} u_n(x, t).$$

The convergence of VIM has been proved in (Abbaoui and Cherruault, 1995).

Example 4.2   With the same fuzzy initial condition has been solved.
One may write Eq. (4.32) as

$$\frac{\partial}{\partial t} \widetilde{y}(t; r) + \frac{\partial^{1-\alpha}}{\partial t^{1-\alpha}} \widetilde{y}(t; r) = 0, \quad 0 < \alpha \le 1, \quad r \in [0, 1] \tag{4.40}$$

According to VIM, we may construct a correction functional as follows:

$$\widetilde{y}_{n+1}(t; r) = \widetilde{y}_n(t; r) + \int_0^t \lambda(\tau) \left\{ \frac{\partial}{\partial \tau} \widetilde{y}(\tau; r) + \frac{\partial^{1-\alpha}}{\partial \tau^{1-\alpha}} \widehat{\widetilde{y}}(\tau; r) \right\} d\tau \tag{4.41}$$

Making the aforementioned correction functional, that is, Eq. (4.41) stationary, and noticing that $\delta \widehat{\widetilde{y}}_n = 0$, we obtain

$$\delta \widetilde{y}_{n+1}(t; r) = \delta \widetilde{y}_n(t; r) + \delta \int_0^t \lambda(\tau) \left\{ \frac{\partial}{\partial \tau} \widetilde{y}(\tau; r) + \frac{\partial^{1-\alpha}}{\partial \tau^{1-\alpha}} \widehat{\widetilde{y}}(\tau; r) \right\} d\tau$$

$$= \delta y_n(t; r) + \lambda(\tau) \delta \widetilde{y}_n(t; r) - \int_0^t \lambda'(\tau) \delta \widetilde{y}_n(\tau; r) d\tau = 0$$

$$= (1 + \lambda(\tau)) \delta \widetilde{y}_n(t; r) - \int_0^t \lambda'(\tau) \delta \widetilde{y}_n(\tau; r) d\tau = 0$$

Thus, we obtain the Euler–Lagrange equation

$$\lambda'(\tau) = 0, \tag{4.42}$$

with natural boundary

$$1 + \lambda(\tau) = 0. \tag{4.43}$$

So, the Lagrange multiplier can be easily identified as

$$\lambda = -1.$$

Substituting the identified Lagrange multiplier into Eq. (4.41), the following variational iteration formula can be obtained:

$$\widetilde{y}_{n+1}(t; r) = \widetilde{y}_n(t; r) - \int_0^t \left\{ \frac{\partial}{\partial \tau} \widetilde{y}(\tau; r) + \frac{\partial^{1-\alpha}}{\partial \tau^{1-\alpha}} \widehat{\widetilde{y}}(\tau; r) \right\} d\tau. \tag{4.44}$$

We start with an initial approximation $\widetilde{y}_0 = \widetilde{y}(0; r)$ given by Eq. (4.36), and by the variational iterational formula (4.44), we have

$$\widetilde{y}_0(t; r) = \left[\underline{y}_0(t; r), \overline{y}_0(t; r)\right] = [0.2r + 0.8, 1.2 - 0.2r],$$

$$\widetilde{y}_1(t; r) = \left[\underline{y}_1(t; r), \overline{y}_1(t; r)\right]$$

$$= \left[(0.2r + 0.8)\left(1 - \frac{t^\alpha}{\Gamma(\alpha + 1)}\right), (1.2 - 0.2r)\left(1 - \frac{t^\alpha}{\Gamma(\alpha + 1)}\right)\right],$$

$$\widetilde{y}_2(t; r) = \left[\underline{y}_2(t; r), \overline{y}_2(t; r)\right]$$

$$= \left[(0.2r + 0.8)\left(1 - \frac{t^\alpha}{\Gamma(\alpha + 1)} + \frac{t^{2\alpha}}{\Gamma(2\alpha + 1)}\right),\right.$$

$$\left.(1.2 - 0.2r)\left(1 - \frac{t^\alpha}{\Gamma(\alpha + 1)} + \frac{t^{2\alpha}}{\Gamma(2\alpha + 1)}\right)\right],$$

$$\widetilde{y}_3(t; r) = \left[\underline{y}_3(t; r), \overline{y}_3(t; r)\right]$$

$$= \left[\begin{array}{c}(0.2r + 0.8)\left(1 - \frac{t^\alpha}{\Gamma(\alpha+1)} + \frac{t^{2\alpha}}{\Gamma(2\alpha+1)} - \frac{t^{3\alpha}}{\Gamma(3\alpha+1)}\right), \\ (1.2 - 0.2r)\left(1 - \frac{t^\alpha}{\Gamma(\alpha+1)} + \frac{t^{2\alpha}}{\Gamma(2\alpha+1)} - \frac{t^{3\alpha}}{\Gamma(3\alpha+1)}\right)\end{array}\right],$$

and so on.

BIBLIOGRAPHY

Abbaoui K, Cherruault Y. Convergence of Adomian's method applied to different equations. *Comput Math Appl* 1994;**28**:103–109.

Abbaoui K, Cherruault Y. New ideas for proving convergence of decomposition methods. *Comput Math Appl* 1995;**29**:103–108.

Adomian G. A new approach to nonlinear partial differential equations. *J Math Anal Appl* 1984;**102**:420–434.

Adomian G. *Solving Frontier problems of Physics: The Decomposition method*. Boston: Kluwer Academic Publishers; 1994.

Cherruault Y. Convergence of Adomian's method. *Kybernetes* 1989;**18**:31–38.

He JH. Homotopy perturbation technique. *Comput Methods Appl Mech Eng* 1999a;**178**:257–262.

He JH. Variational iteration method kind of non-linear analytical technique: Some examples. *Int J Non-Linear Mech* 1999b;**34**:699–708.

He JH. A coupling method of homotopy technique and a perturbation technique for nonlinear problems. *Int J Non-linear Mech* 2000a;**35**:37–43.

He JH. Variational iteration method for autonomous ordinary differential system. *Appl Math Comput* 2000b;**114**:115–123.

Himoun N, Abbaoui K, Cherruault Y. New results of convergence of Adomian's method. *Kybernetes* 1999;**28**:423–429.

Kazem S. Exact solution of some linear fractional differential equations by Laplace transform. *Int J Nonlinear Sci* 2013;**16**:3–11.

Tapaswini S, Chakraverty S. Numerical solution of n-th order fuzzy linear differential equations by homotopy perturbation method. *Int J Comput Appl* 2013;**64**:5–10.

5

FUZZY FRACTIONAL HEAT EQUATIONS

The target of this chapter is to investigate the solution of fuzzy arbitrary-order heat equations using homotopy perturbation method (HPM). Initial conditions are taken in terms of triangular fuzzy number.

5.1 ARBITRARY-ORDER HEAT EQUATION

Consider the fuzzy arbitrary-order heat equation with the indicated initial conditions (Salah et al., 2013)

$$\frac{\partial^\alpha \widetilde{U}}{\partial t^\alpha} = \frac{\partial^2 \widetilde{U}}{\partial x^2} + \widetilde{k}, \quad 0 \le x \le 1, \ t > 0,$$

where,

$$\widetilde{U}(x,0) = \widetilde{f}(x), \quad 0 \le x \le 1$$

5.2 SOLUTION OF FUZZY ARBITRARY-ORDER HEAT EQUATIONS BY HPM

Let us consider the fuzzy arbitrary-order heat equations

$$\frac{\partial^\alpha \widetilde{U}}{\partial t^\alpha} = \frac{\partial^2 \widetilde{U}}{\partial x^2} + \widetilde{k} \tag{5.1}$$

Fuzzy Arbitrary Order System: Fuzzy Fractional Differential Equations and Applications, First Edition.
Snehashish Chakraverty, Smita Tapaswini, and Diptiranjan Behera.
© 2016 John Wiley & Sons, Inc. Published 2016 by John Wiley & Sons, Inc.

with fuzzy initial conditions

$$\widetilde{U}(x, 0) = \widetilde{f}(x).$$

Using Hukahara differentiability, the fuzzy arbitrary-order heat equations that is, Eq. (5.1) may be reduced to a set of ordinary differential equations.

$$\frac{\partial^\alpha \underline{U}}{\partial t^\alpha} = \frac{\partial^2 \underline{U}}{\partial x^2} + \underline{k},$$

$$\frac{\partial^\alpha \overline{U}}{\partial t^\alpha} = \frac{\partial^2 \overline{U}}{\partial x^2} + \overline{k},$$

with fuzzy initial conditions

$$\left[\underline{U}(x, 0), \overline{U}(x, 0)\right] = \left[\underline{f}(x), \overline{f}(x)\right].$$

According to HPM, we may construct a simple homotopy for an embedding parameter $p \in [0, 1]$ as follows:

$$(1 - p)\frac{\partial^\alpha \widetilde{U}}{\partial t^\alpha} + p\left[\frac{\partial^\alpha \widetilde{U}}{\partial t^\alpha} - \frac{\partial^2 \widetilde{U}}{\partial x^2} - \widetilde{k}\right] = 0 \qquad (5.2)$$

or

$$\frac{\partial^\alpha \widetilde{U}}{\partial t^\alpha} + p\left[-\frac{\partial^2 \widetilde{U}}{\partial x^2} - \widetilde{k}\right] = 0. \qquad (5.3)$$

Here, p is considered as a small homotopy parameter $0 \leq p \leq 1$. For $p = 0$, Eqs. (5.2) and (5.3) become a linear equation, that is, $\partial^\alpha \widetilde{U}/\partial t^\alpha = 0$, which is easy to solve. For $p = 1$, Eqs. (5.2) and (5.3) turn out to be the same as the original equation (5.1). This is called deformation in topology. $\partial^\alpha \widetilde{U}/\partial t^\alpha$ and $-\frac{\partial^2 \widetilde{U}}{\partial x^2} - \widetilde{k}$ are called homotopic. Next, we can assume the solution of Eq. (5.2) or (5.3) as a power series expansion in p as

$$\widetilde{U}(t) = \widetilde{U}_0(t) + p\widetilde{U}_1(t) + p^2\widetilde{U}_2(t) + \cdots, \qquad (5.4)$$

where $\widetilde{U}_i(t)$, $i = 0, 1, 2, \ldots$ are functions yet to be determined. Substituting Eq. (5.4) into Eq. (5.2) or (5.3), and equating the terms with the identical powers of p, we have

$$p^0 : \frac{\partial^\alpha \widetilde{U}_0}{\partial t^\alpha} = 0,$$

$$p^1 : \frac{\partial^\alpha \widetilde{U}_1}{\partial t^\alpha} - \frac{\partial^2 \widetilde{U}_0}{\partial x^2} - \widetilde{k} = 0, \qquad (5.5)$$

$$p^2 : \frac{\partial^\alpha \widetilde{U}_2}{\partial t^\alpha} - \frac{\partial^2 \widetilde{U}_1}{\partial x^2} = 0,$$

$$p^3 : \frac{\partial^\alpha \widetilde{U}_3}{\partial t^\alpha} - \frac{\partial^2 \widetilde{U}_2}{\partial x^2} = 0,$$

so on.

The method is based on applying the operator J_t^α (the inverse operator of Caputo derivative D_t^α) on both sides of Eq. (5.5), then we can calculate the approximate solution bounds $\widetilde{U}(t) = \sum_{n=0}^\infty \widetilde{U}_n(t)$ term by term. Otherwise, an approximation to the solutions would be achieved by computing some terms, say k, as

$$\widetilde{U}(t) = \lim_{k\to\infty} \underline{\xi}_k(t), \tag{5.6}$$

where

$$\underline{\xi}_k(t) = \sum_{n=0}^k \widetilde{U}_n(t).$$

5.3 NUMERICAL EXAMPLES

Example 5.1 Let us consider the following fuzzy arbitrary-order heat equation (Salah et al., 2013):

$$\frac{\partial^\alpha \widetilde{U}}{\partial t^\alpha} = \frac{\partial^2 \widetilde{U}}{\partial x^2}, \quad 0 < x < 1, \ t > 0,$$

with initial condition $\widetilde{U}(x,0) = \widetilde{f}(x) = \widetilde{k} \sin(\pi x)$, $0 < x < 1$, where $\widetilde{k} = [\underline{k}(r), \overline{k}(r)] = [r - 1, \ 1 - r]$.

Solving Eq. (5.5), one may have

$$\underline{U}_0(x, t; r) = \underline{k}(r) \sin(\pi x),$$

$$\overline{U}_0(x, t; r) = \overline{k}(r) \sin(\pi x),$$

$$\underline{U}_1(x, t; r) = -\underline{k}(r)\pi^2 \sin(\pi x) \frac{t^\alpha}{\Gamma(\alpha + 1)},$$

$$\overline{U}_1(x, t; r) = -\overline{k}(r)\pi^2 \sin(\pi x) \frac{t^\alpha}{\Gamma(\alpha + 1)},$$

$$\underline{U}_2(x, t; r) = \underline{k}(r)\pi^4 \sin(\pi x) \frac{t^{2\alpha}}{\Gamma(2\alpha + 1)},$$

$$\overline{U}_2(x, t; r) = \overline{k}(r)\pi^4 \sin(\pi x) \frac{t^{2\alpha}}{\Gamma(2\alpha + 1)},$$

$$\underline{U}_3(x, t; r) = -\underline{k}(r)\pi^6 \sin(\pi x) \frac{t^{3\alpha}}{\Gamma(3\alpha + 1)},$$

$$\overline{U}_3(x, t; r) = -\overline{k}(r)\pi^6 \sin(\pi x) \frac{t^{3\alpha}}{\Gamma(3\alpha + 1)}.$$

In a similar manner, the higher approximation may be obtained as discussed earlier. Substituting these in Eq. (5.4), we may get the approximate solution of $\widetilde{U}(x, t)$. Accordingly, the exact solution is

$$\underline{U}(x, t; r) = \sum_{m=0}^\infty \frac{(-1)^m \pi^{2m} t^{m\alpha}}{\Gamma(m\alpha + 1)} \underline{k}(r) \sin(\pi x),$$

$$\overline{U}(x,t;r) = \sum_{m=0}^{\infty} \frac{(-1)^m \pi^{2\alpha} t^{m\alpha}}{\Gamma(m\alpha + 1)} \overline{k}(r) \sin(\pi x).$$

$$\widetilde{U}(x,t;r) = \left[\sum_{m=0}^{\infty} \frac{(-1)^m \pi^{2m} t^{m\alpha}}{\Gamma(m\alpha + 1)} \underline{k}(r) \sin(\pi x), \sum_{m=0}^{\infty} \frac{(-1)^m \pi^{2m} t^{m\alpha}}{\Gamma(m\alpha + 1)} \overline{k}(r) \sin(\pi x)\right].$$

Example 5.2 Next, consider another fuzzy arbitrary-order heat equation (Salah et al., 2013).

$$\frac{\partial^\alpha \widetilde{U}}{\partial t^\alpha} = \frac{1}{2} x^2 \frac{\partial^2 \widetilde{U}}{\partial x^2}, \quad 0 < x < 1, \quad t > 0,$$

with the initial condition $\widetilde{U}(x,0) = \widetilde{f}(x) = \widetilde{k}x^2$, $0 < x < 1$, where $\widetilde{k} = [\underline{k}(r), \overline{k}(r)] = [r - 1, 1 - r]$.

Applying HPM in this case, one may get

$$\underline{U}_0(x,t;r) = \underline{k}(r)x^2,$$

$$\overline{U}_0(x,t;r) = \overline{k}(r)x^2,$$

$$\underline{U}_1(x,t;r) = \underline{k}(r)x^2 \frac{t^\alpha}{\Gamma(\alpha + 1)},$$

$$\overline{U}_1(x,t;r) = \overline{k}(r)x^2 \frac{t^\alpha}{\Gamma(\alpha + 1)},$$

$$\underline{U}_2(x,t;r) = \underline{k}(r)x^2 \frac{t^{2\alpha}}{\Gamma(2\alpha + 1)},$$

$$\overline{U}_2(x,t;r) = \overline{k}(r)x^2 \frac{t^{2\alpha}}{\Gamma(2\alpha + 1)},$$

$$\underline{U}_3(x,t;r) = \underline{k}(r)x^2 \frac{t^{3\alpha}}{\Gamma(3\alpha + 1)},$$

$$\overline{U}_3(x,t;r) = \overline{k}(r)x^2 \frac{t^{3\alpha}}{\Gamma(3\alpha + 1)}.$$

In a similar manner, higher approximations may be obtained as discussed earlier. Substituting these in Eq. (4.15), one may get the approximate solution of $\widetilde{U}(x,t)$. Accordingly, the exact solution is

$$\underline{U}(x,t;r) = \sum_{m=0}^{\infty} \frac{t^{m\alpha}}{\Gamma(m\alpha + 1)} \underline{k}(r)x^2 = \underline{k}(r)E_\alpha(t)x^2,$$

$$\overline{U}(x,t;r) = \sum_{m=0}^{\infty} \frac{t^{m\alpha}}{\Gamma(m\alpha + 1)} \overline{k}(r)x^2 = \overline{k}(r)E_\alpha(t)x^2,$$

$$\widetilde{U}(x,t;r) = \left[\underline{k}(r)E_\alpha(t)x^2, \overline{k}(r)E_\alpha(t)x^2\right].$$

$E_\alpha(t)$ is called the Mittag-Leffler function of one parameter where,

$$E_\alpha(t) = \sum_{m=0}^{\infty} \frac{t^m}{\Gamma(\alpha m + 1)},$$

5.4 NUMERICAL RESULTS

In this section, the numerical results of fuzzy arbitrary (fractional)-order heat equations using HPM have been presented. Few results as per the following cases are reported here.

In the first example, fuzzy arbitrary-order heat equation with fuzzy initial condition as discussed earlier is solved by HPM. Fuzzy results are depicted in Fig. 5.1(a)–(c)

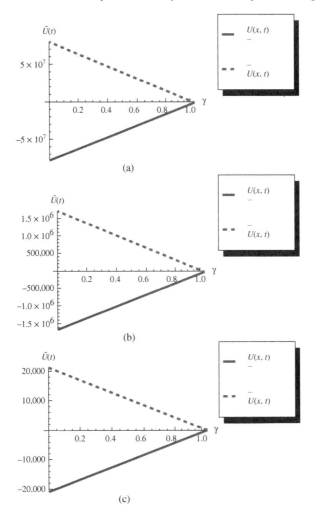

Figure 5.1 Solution of fuzzy arbitrary (fractional)-order heat equations of Example 4.1 for $\tilde{U}(x, t)$ at $x = t = 0.5$, and $r \in [0, 1]$ for (a) $\alpha = 1/3$, (2) $\alpha = 1/2$, and (c) $\alpha = 2/3$

for various arbitrary-order derivatives, namely $\alpha = \beta = 1/3$, $\alpha = \beta = 1/2$, and $\alpha = \beta = 2/3$ and $x = t = 0.5$, respectively.

Similar types of studies have also been conducted for Example 5.2. The corresponding fuzzy results are given in Fig. 5.2(a)–(c) for various arbitrary-order (similarly to Example 5.1) derivatives, namely $\alpha = \beta = 1/3$, $\alpha = \beta = 1/2$ and $\alpha = \beta = 2/3$, and $x = t = 0.5$, respectively.

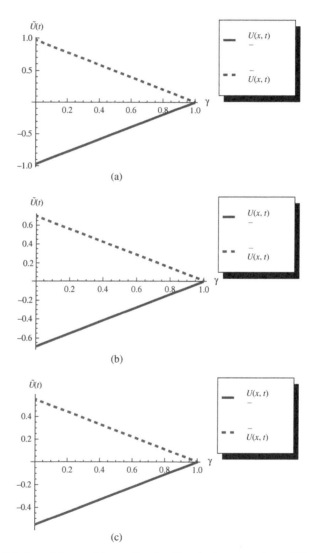

Figure 5.2 Solution of fuzzy arbitrary (fractional)-order heat equations of Example 5.2 for $\widetilde{U}(x, t)$ at $x = t = 0.5$ and $\gamma \in [0, 1]$ for (a) $\alpha = 1/3$ (b) $\alpha = 1/2$, and (c) $\alpha = 2/3$

It may be worth mentioning that the obtained results for fuzzy arbitrary (fractional)-order heat equations by HPM exactly agree with the solution of fuzzy fractional-order heat equations presented by Salah et al. (2013) using homotopy analysis transform method.

Here, HPM has been applied to find the solution of fuzzy arbitrary-order heat equations. The solution obtained by HPM is an infinite series with appropriate initial condition, which is expressed in a closed form giving the exact solution. The result shows that the HPM is a powerful tool to fuzzy arbitrary-order heat equations.

BIBLIOGRAPHY

Agrawal RP, Lakshmikantham V, Nieto JJ. On the concept of solution for fractional differential equations with uncertainty. *Nonlinear Anal.* 2010;**72**:2859–2862.

Allahviranloo T, Salahshour S, Abbasbandy S. Explicit solutions of fractional differential equations with uncertainty. *Soft Comput* 2012;**16**:297–302.

Arshad S, Lupulescu V. On the fractional differential equations with uncertainty. *Nonlinear Anal* 2011a;**74**:3685–3693.

Arshad S, Lupulescu V. Fractional differential equation with the fuzzy initial condition. *Electron J Differ Eqs* 2011b;**2011**:1–8.

Chakraverty S. *Mathematics of Uncertainty Modeling in the Analysis of Engineering and Science Problems*. USA: IGI Global Publication; 2014.

Dubois D, Prade H. *Fuzzy Sets and systems: Theory and Applications*. New York: Academic Press; 1980.

Hanss M. *Applied Fuzzy Arithmetic: An Introduction with Engineering Applications*. Berlin: Springer-Verlag; 2005.

He JH. Homotopy perturbation technique. *Comput Math Appl Mech Eng* 1999;**178**:257–262.

He JH. A coupling method of a homotopy technique and a perturbation technique for non-linear problems. *Int J Nonlinear Mech* 2000;**35**:37–43.

Kiryakova VS. *Generalized Fractional Calculus and Applications*. Harlow, England: Longman Scientific & Technical, Longman House, Burnt Mill; 1993.

Miller KS, Ross B. *An Introduction to the Fractional Calculus and Fractional Differential Equations*. New York, NY: John Wiley & Sons, Inc.; 1993.

Mohammed OH, Fadhel SF, Abdul-Khaleq FA. Differential transform method for solving fuzzy fractional initial value problems. *J Basrah Res (Sci)* 2011;**37**:158–170.

Oldham KB, Spanier J. *The Fractional Calculus*. New York, NY: Academic Press; 1974.

Petras I. *Fractional-Order Nonlinear Systems Modeling, Analysis and Simulation*. London, New York: Higer Education Press, Springer Heidelberg Dordrecht; 2011.

Podlubny I. *Fractional Differential Equations*. New York, NY: Academic Press; 1999.

Ross TJ. *Fuzzy Logic with Engineering Applications*. New York: John Wiley & Sons; 2004.

Salah A, Khan M, Gondal AM. A novel solution procedure for fuzzy fractional heat equations by homotopy analysis transform method. *Neural Comput Appl* 2013;**23**:269–271.

Salahshour S, Allahviranloo T, Abbasbandy S. Solving fuzzy fractional differential equations by fuzzy Laplace transforms. *Commun Nonlinear Sci Numer Simul* 2012;**17**:1372–1381.

Samko SG, Kilbas AA, Marichev OI. *Fractional Integrals and Derivatives-Theory and Applications*. Langhorne, PA: Gordon and Breach Science Publishers; 1993.

Smita T, Chakraverty S. Numerical solution of n-th order fuzzy linear differential equations by homotopy perturbation method. *Int J Comput Appl* 2013;**64**:5–10.

Wang H, Liu Y. Existence results for fractional fuzzy differential equations with finite delay. *Int Math Forum* 2011;**6**:2535–2538.

Zimmermann HJ. *Fuzzy Set Theory and Its Application*. Boston/Dordrecht/London: Kluwer Academic Publishers; 2001.

6

FUZZY FRACTIONAL BIOMATHEMATICAL APPLICATIONS

In this chapter, numerical solution of fuzzy arbitrary/fractional-order predator–prey is investigated. In the predator–prey equation, fuzziness appeared in the initial conditions, which are taken in terms of triangular fuzzy number, and HPM is used to solve the problem.

6.1 FUZZY ARBITRARY-ORDER PREDATOR–PREY EQUATIONS

Let us consider the fuzzy arbitrary-order predator–prey equations (Tapaswini and Chakraverty, 2013)

$$D_t^\alpha \widetilde{x}(t) = a(t)\widetilde{x}(t) - b(t)\widetilde{x}(t)\widetilde{y}(t),$$

$$D_t^\lambda \widetilde{y}(t) = c(t)\widetilde{x}(t)\widetilde{y}(t) - d(t)\widetilde{y}(t), \qquad (6.1)$$

with triangular fuzzy initial conditions

$$\widetilde{x}(0) = (a_1, b_1, c_1) \quad \text{and} \quad \widetilde{y}(0) = (a_2, b_2, c_2).$$

Using r-cut, triangular fuzzy initial conditions become

$$\widetilde{x}(0; r) = [r(b_1 - a_1) + a_1, c_1 - r(c_1 - b_1)]$$

Fuzzy Arbitrary Order System: Fuzzy Fractional Differential Equations and Applications, First Edition.
Snehashish Chakraverty, Smita Tapaswini, and Diptiranjan Behera.
© 2016 John Wiley & Sons, Inc. Published 2016 by John Wiley & Sons, Inc.

and

$$\widetilde{y}(0; r) = [r(b_2 - a_2) + a_2, c_2 - r(c_2 - b_2)], \quad 0 \le r \le 1.$$

Here, $\widetilde{x}(t; r) = [\underline{x}(t; r), \overline{x}(t; r)]$ and $\widetilde{y}(t; r) = [\underline{y}(t; r), \overline{y}(t; r)]$ are fuzzy functions of t. Using Hukahara differentiability, fuzzy arbitrary-order predator–prey equation (Eq. (6.1)) may be reduced to a set of ordinary differential equations. One can readily construct homotopy for Eq. (6.1) as

$$(1 - p)D_t^\alpha \widetilde{x}(t; r) + p[D_t^\alpha \widetilde{x}(t; r) - a(t)\widetilde{x}(t; r) + b(t)\widetilde{x}(t; r)\widetilde{y}(t; r)] = 0, \tag{6.2}$$

$$(1 - p)D_t^\lambda \widetilde{y}(t; r) + p[D_t^\lambda \widetilde{y}(t; r) - c(t)\widetilde{x}(t; r)\widetilde{y}(t; r) - d(t)\widetilde{y}(t; r)] = 0. \tag{6.3}$$

One may try to obtain solutions of Eqs. (6.2) and (6.3) in the form

$$\widetilde{x}(t; r) = \widetilde{x}_0(t; r) + p\widetilde{x}_1(t; r) + p^2\widetilde{x}_2(t; r) + \cdots, \tag{6.4}$$

$$\widetilde{y}(t; r) = \widetilde{y}_0(t; r) + p\widetilde{y}_1(t; r) + p^2\widetilde{y}_2(t; r) + \cdots, \tag{6.5}$$

where $\widetilde{x}_i(t; r)$, $i = 0, 1, 2, \dots$ and $\widetilde{y}_i(t; r)$, $i = 0, 1, 2, \dots$ are functions yet to be determined. Substituting Eqs. (6.4) and (6.5) in Eqs. (6.2) and (6.3), respectively, and equating the terms with the identical powers of p, we have

$$p^0 : \begin{cases} D_t^\alpha \widetilde{x}_0(t; r) = 0, \\ D_t^\lambda \widetilde{y}_0(t; r) = 0, \end{cases}$$

$$p^1 : \begin{cases} D_t^\alpha \widetilde{x}_1(t; r) = a(t)\widetilde{x}_0(t; r) - b(t)\widetilde{x}_0(t; r)\widetilde{y}_0(t; r), \\ D_t^\lambda \widetilde{y}_1(t; r) = c(t)\widetilde{x}_0(t; r)\widetilde{y}_0(t; r) - d(t)\widetilde{y}_0(t; r), \end{cases}$$

$$p^2 : \begin{cases} D_t^\alpha \widetilde{x}_2(t; r) = a(t)\widetilde{x}_1(t; r) - b(t)\widetilde{x}_1(t; r)\widetilde{y}_0(t; r) - b(t)\widetilde{x}_0(t; r)\widetilde{y}_1(t; r), \\ D_t^\lambda \widetilde{y}_2(t; r) = c(t)\widetilde{x}_0(t; r)\widetilde{y}_1(t; r) + c(t)\widetilde{x}_1(t; r)\widetilde{y}_0(t; r) - d(t)\widetilde{y}_1(t; r), \end{cases}$$

$$p^3 : \begin{cases} D_t^\alpha \widetilde{x}_3(t; r) = a(t)\widetilde{x}_2(t; r) - b(t)\widetilde{x}_2(t; r)\widetilde{y}_0(t) - b(t)\widetilde{x}_1(t)\widetilde{y}_1(t; r) \\ \quad -b(t)\widetilde{x}_0(t; r)\widetilde{y}_2(t; r), \\ D_t^\lambda \widetilde{y}_3(t; r) = c(t)\widetilde{x}_2(t; r)\widetilde{y}_0(t; r) + c(t)\widetilde{x}_1(t; r)\widetilde{y}_1(t; r) + c(t)\widetilde{x}_0(t; r)\widetilde{y}_2(t; r) \\ \quad -d(t)\widetilde{y}_2(t; r), \end{cases}$$

$$\tag{6.6}$$

and so on.

The method is based on applying the operators J_t^α and J_t^λ (the inverse operator of Caputo derivatives D_t^α and D_t^λ, respectively) on both sides of each of the equations (6.6) and then the approximate solution bounds can be written as

$$\widetilde{x}(t; r) = \left[\sum_{n=0}^\infty \underline{x}_n(t; r), \sum_{n=0}^\infty \overline{x}_n(t; r) \right] \text{ and } \widetilde{y}(t; r) = \left[\sum_{n=0}^\infty \underline{y}_n(t; r), \sum_{n=0}^\infty \overline{y}_n(t; r) \right].$$

An approximation to the solutions would be achieved by computing few terms, say k, as

$$\underline{x}(t;r) = \lim_{k \to \infty} \underline{\xi}_k(t;r) \text{ and } \underline{y}(t;r) = \lim_{k \to \infty} \underline{\phi}_k(t;r),$$

$$\overline{x}(t;r) = \lim_{k \to \infty} \overline{\xi}_k(t;r) \text{ and } \overline{y}(t;r) = \lim_{k \to \infty} \overline{\phi}_k(t;r). \tag{6.7}$$

Here

$$\underline{\xi}_k(t;r) = \sum_{n=0}^{k} \underline{x}_n(t;r), \ \underline{\phi}_k(t;r) = \sum_{n=0}^{k} \underline{y}_n(t;r), \ \overline{\xi}_k(t;r) = \sum_{n=0}^{k} \overline{x}_n(t;r) \text{ and } \overline{\phi}_k(t;r)$$

$$= \sum_{n=0}^{k} \overline{y}_n(t;r).$$

6.1.1 Particular Case

Let us consider $a(t) = t$, $b(t) = 1$, $c(t) = 1$, $d(t) = t$, and initial conditions in terms of triangular fuzzy numbers, namely $\widetilde{x}(0) = (1.1, 1.3, 1.5)$ and $\widetilde{y}(0) = (0.4, 0.6, 0.8)$.

Through r-cut approach, fuzzy initial conditions become

$$\widetilde{x}(0;r) = [0.2r + 1.1, 1.5 - 0.2r] = [\delta_1, \delta_2]$$

and

$$\widetilde{y}(0;r) = [0.2r + 0.4, 0.8 - 0.2r] = [\eta_1, \eta_2].$$

Solving Eq. (6.6), one may have

$$\underline{x}_0(t;r) = \delta_1, \quad \overline{x}_0(t;r) = \delta_2,$$

$$\underline{y}_0(t;r) = \eta_1, \quad \overline{y}_0(t;r) = \eta_2,$$

$$\underline{x}_1(t;r) = \frac{t^{\alpha+1}}{\Gamma(\alpha+2)}\delta_1 - \frac{t^\alpha}{\Gamma(\alpha+1)}\delta_2\eta_2, \quad \overline{x}_1(t;r) = \frac{t^{\alpha+1}}{\Gamma(\alpha+2)}\delta_2 - \frac{t^\alpha}{\Gamma(\alpha+1)}\delta_1\eta_1,$$

$$\overline{y}_1(t;r) = \frac{t^\lambda}{\Gamma(\lambda+1)}\delta_2\eta_2 - \frac{t^{\lambda+1}}{\Gamma(\lambda+2)}\eta_1,$$

$$\underline{x}_2(t;r) = \frac{t^{2\alpha}}{\Gamma(2\alpha+1)}\delta_1\eta_1\eta_2 - \frac{(\alpha+2)t^{2\alpha+1}}{\Gamma(2\alpha+2)}\delta_2\eta_2 + \frac{(\alpha+2)t^{2\alpha+2}}{\Gamma(2\alpha+3)}\delta_1$$

$$- \frac{t^{\alpha+\lambda}}{\Gamma(\alpha+\lambda+1)}\delta_2{}^2\eta_2 + \frac{t^{\alpha+\lambda+1}}{\Gamma(\alpha+\lambda+2)}\eta_1\delta_2,$$

$$\overline{x}_2(t;r) = \frac{t^{2\alpha}}{\Gamma(2\alpha+1)}\delta_2\eta_2\eta_1 - \frac{(\alpha+2)t^{2\alpha+1}}{\Gamma(2\alpha+2)}\delta_1\eta_1 + \frac{(\alpha+2)t^{2\alpha+2}}{\Gamma(2\alpha+3)}\delta_2$$

$$- \frac{t^{\alpha+\lambda}}{\Gamma(\alpha+\lambda+1)}\delta_1{}^2\eta_1 + \frac{t^{\alpha+\lambda+1}}{\Gamma(\alpha+\lambda+2)}\eta_2\delta_1,$$

$$\underline{y}_2(t;r) = \frac{t^{2\lambda}}{\Gamma(\lambda+1)}\delta_1{}^2\eta_1 - \frac{t^{2\lambda+1}}{\Gamma(2\lambda+2)}\eta_2\{\delta_1+(\lambda+1)\delta_2\} + \frac{(\lambda+2)t^{2\lambda+2}}{\Gamma(2\lambda+3)}\eta_1$$

$$- \frac{t^{\alpha+\lambda}}{\Gamma(\alpha+\lambda+1)}\delta_2\eta_2\eta_1 + \frac{t^{\alpha+\lambda+1}}{\Gamma(\alpha+\lambda+2)}\delta_1\eta_1,$$

$$\bar{y}_2(t;r) = \frac{t^{2\lambda}}{\Gamma(\lambda+1)}\delta_2{}^2\eta_2 - \frac{t^{2\lambda+1}}{\Gamma(2\lambda+2)}\eta_1\{\delta_2+(\lambda+1)\delta_1\} + \frac{(\lambda+2)t^{2\lambda+2}}{\Gamma(2\lambda+3)}\eta_2$$

$$- \frac{t^{\alpha+\lambda}}{\Gamma(\alpha+\lambda+1)}\delta_1\eta_1\eta_2 + \frac{t^{\alpha+\lambda+1}}{\Gamma(\alpha+\lambda+2)}\delta_2\eta_1,$$

$$\underline{x}_3(t;r) = -\frac{t^{3\alpha}}{\Gamma(3\alpha+1)}\eta_2{}^2\delta_2\eta_1 + \frac{3(\gamma+1)t^{3\alpha+1}}{\Gamma(3\alpha+2)}\eta_1\delta_1\eta_2 - \frac{(\alpha+2)(2\alpha+3)t^{3\alpha+2}}{\Gamma(3\alpha+3)}\delta_2\eta_2$$

$$+ \frac{(\alpha+2)(2\alpha+3)t^{3\alpha+3}}{\Gamma(3\alpha+4)}\delta_1 - \frac{t^{\alpha+2\lambda}}{\Gamma(\alpha+2\lambda+1)}\delta_2{}^3\eta_2$$

$$+ \frac{t^{\alpha+2\lambda+1}}{\Gamma(\alpha+2\lambda+2)}\delta_2\eta_1\{\delta_2+(\lambda+1)\delta_1\} - \frac{(\lambda+2)t^{\alpha+2\lambda+2}}{\Gamma(\alpha+2\lambda+3)}\delta_2\eta_2$$

$$+ \frac{t^{2\alpha+\lambda}}{\Gamma(2\alpha+\lambda+1)}\eta_2\eta_1\delta_1 - \frac{t^{2\alpha+\lambda+1}}{\Gamma(2\alpha+\lambda+2)}\xi_1 + \frac{t^{2\alpha+\lambda+2}}{\Gamma(2\alpha+\lambda+3)}\eta_1\delta_2\psi_1,$$

where

$$\xi_1 = (\alpha+\lambda+2)\delta_2{}^2\eta_2 + \eta_2{}^2\delta_1 + \frac{\Gamma(\alpha+\lambda+2)}{\Gamma(\alpha+2)\Gamma(\lambda+1)}(\delta_2{}^2\eta_2+\delta_1\eta_1{}^2)$$

and

$$\psi_1 = (\alpha+\lambda+2) + \frac{\Gamma(\alpha+\lambda+3)}{\Gamma(\alpha+2)\Gamma(\lambda+2)}.$$

$$\bar{x}_3(t;r) = -\frac{t^{3\alpha}}{\Gamma(3\alpha+1)}\eta_1{}^2\delta_1\eta_2 + \frac{3(\gamma+1)t^{3\alpha+1}}{\Gamma(3\alpha+2)}\eta_2\delta_2\eta_1 - \frac{(\alpha+2)(2\alpha+3)t^{3\alpha+2}}{\Gamma(3\alpha+3)}\delta_1\eta_1$$

$$+ \frac{(\alpha+2)(2\alpha+3)t^{3\alpha+3}}{\Gamma(3\alpha+4)}\delta_2\frac{t^{\alpha+2\lambda}}{\Gamma(\alpha+2\lambda+1)}\delta_1{}^3\eta_1$$

$$+ \frac{t^{\alpha+2\lambda+1}}{\Gamma(\alpha+2\lambda+2)}\delta_1\eta_2\{\delta_1+(\lambda+1)\delta_2\} - \frac{(\lambda+2)t^{\alpha+2\lambda+2}}{\Gamma(\alpha+2\lambda+3)}\delta_1\eta_1$$

$$+ \frac{t^{2\alpha+\lambda}}{\Gamma(2\alpha+\lambda+1)}\eta_1\eta_2\delta_2\left\{\delta_2 + \frac{\Gamma(\alpha+\lambda+1)}{\Gamma(\alpha+1)\Gamma(\lambda+1)}\delta_1 + \delta_1\right\}$$

$$- \frac{t^{2\alpha+\lambda+1}}{\Gamma(2\alpha+\lambda+2)}\xi_2 + \frac{t^{2\alpha+\lambda+2}}{\Gamma(2\alpha+\lambda+3)}\eta_2\delta_1\psi_2,$$

where

$$\xi_2 = (\alpha + \lambda + 2)\delta_1{}^2\eta_1 + \eta_1{}^2\delta_2 + \frac{\Gamma(\alpha + \lambda + 2)}{\Gamma(\alpha + 2)\Gamma(\lambda + 1)}(\delta_1{}^2\eta_1 + \delta_2\eta_2{}^2)$$

and

$$\psi_2 = (\alpha + \lambda + 2) + \frac{\Gamma(\alpha + \lambda + 3)}{\Gamma(\alpha + 2)\Gamma(\lambda + 2)}.$$

$$\begin{aligned}
\underline{y}_3(t; r) = {}& \frac{t^{3\beta}}{\Gamma(3\beta + 1)}\delta_1{}^3\eta_1 - \frac{t^{3\beta+1}}{\Gamma(3\beta + 2)} \\
& \times \{(2\beta + 1)\delta_2{}^2\eta_2 + \delta_1{}^2\eta_2 + (\beta + 1)\delta_2\eta_2\delta_1\} \\
& + \frac{t^{3\beta+2}}{\Gamma(3\beta + 3)}\eta_1\{(\beta + 2)\delta_1 + 2(\beta + 1)\delta_2 + 2(\beta + 1)^2\delta_1\} \\
& + \frac{(\beta + 2)(2\alpha + 3)t^{3\beta+3}}{\Gamma(3\beta + 4)}\eta_2 + \frac{t^{2\alpha+\beta}}{\Gamma(2\alpha + \beta + 1)}\delta_1\eta_1{}^2\eta_2 \\
& - \frac{(\alpha + 2)t^{2\alpha+\beta+1}}{\Gamma(2\alpha + \beta + 2)}\delta_2\eta_1\eta_2 + \frac{(\alpha + 2)t^{2\alpha+\beta+2}}{\Gamma(2\alpha + \beta + 3)}\delta_1\eta_1 \\
& + \frac{t^{\alpha+2\beta+1}}{\Gamma(\alpha + 2\beta + 2)}\sigma_1 - \frac{t^{\alpha+2\beta+2}}{\Gamma(\alpha + 2\beta + 3)}\eta_2\rho_1, \\
& - \frac{t^{\alpha+2\beta}}{\Gamma(\alpha + 2\beta + 1)}\eta_1\delta_2\eta_2\left\{\delta_1 + \frac{\Gamma(\alpha + \beta + 1)}{\Gamma(\alpha + 1)\Gamma(\beta + 1)}\delta_1 + \delta_2\right\} \\
& + \frac{t^{\alpha+2\beta+1}}{\Gamma(\alpha + 2\beta + 2)}\sigma_1 - \frac{t^{\alpha+2\beta+2}}{\Gamma(\alpha + 2\beta + 3)}\eta_2\rho_1,
\end{aligned}$$

where

$$\begin{aligned}
\sigma_1 = {}& \delta_1{}^2\eta_1\left(1 + \frac{\Gamma(\alpha + \lambda + 2)}{\Gamma(\alpha + 2)\Gamma(\lambda + 1)}\right) + \frac{\Gamma(\alpha + \lambda + 2)}{\Gamma(\alpha + 1)\Gamma(\lambda + 2)}\delta_2\eta_2{}^2 \\
& + \eta_1{}^2\delta_2 + (\alpha + \lambda + 1)\delta_1\eta_2\eta_1
\end{aligned}$$

and

$$\rho_1 = (\alpha + \lambda + 2)\delta_2 + \frac{\Gamma(\alpha + \lambda + 3)}{\Gamma(\alpha + 2)\Gamma(\lambda + 2)}\delta_1\eta_2.$$

$$\bar{y}_3(t;r) = \frac{t^{3\beta}}{\Gamma(3\beta+1)}\delta_2{}^3\eta_2 - \frac{t^{3\beta+1}}{\Gamma(3\beta+2)}\{(2\beta+1)\delta_1{}^2\eta_1 + \delta_2{}^2\eta_1 + (\beta+1)\delta_1\eta_1\delta_2\}$$

$$+ \frac{t^{3\beta+2}}{\Gamma(3\beta+3)}\eta_2\{(\beta+2)\delta_2 + 2(\beta+1)\delta_1 + 2(\beta+1)^2\delta_2\}$$

$$+ \frac{(\beta+2)(2\alpha+3)t^{3\beta+3}}{\Gamma(3\beta+4)}\eta_1 + \frac{t^{2\alpha+\beta}}{\Gamma(2\alpha+\beta+1)}\delta_2\eta_2{}^2\eta_1$$

$$- \frac{(\alpha+2)t^{2\alpha+\beta+1}}{\Gamma(2\alpha+\beta+2)}\delta_1\eta_1\eta_2 + \frac{(\alpha+2)t^{2\alpha+\beta+2}}{\Gamma(2\alpha+\beta+3)}\delta_2\eta_2 - \frac{t^{\alpha+2\beta}}{\Gamma(\alpha+2\beta+1)}\eta_2\delta_1\eta_1$$

$$\times \left\{\delta_2 + \frac{\Gamma(\alpha+\beta+1)}{\Gamma(\alpha+1)\Gamma(\beta+1)}\delta_2 + \delta_1\right\} + \frac{t^{\alpha+2\beta+1}}{\Gamma(\alpha+2\beta+2)}\sigma_2$$

$$- \frac{t^{\alpha+2\beta+2}}{\Gamma(\alpha+2\beta+3)}\eta_1\rho_2 \,,$$

where

$$\sigma_2 = \delta_2{}^2\eta_2\left(1 + \frac{\Gamma(\alpha+\lambda+2)}{\Gamma(\alpha+2)\Gamma(\lambda+1)}\right)$$

$$+ \frac{\Gamma(\alpha+\lambda+2)}{\Gamma(\alpha+1)\Gamma(\nu+2)}\delta_1\eta_1{}^2 + \eta_2{}^2\delta_1 + (\alpha+\lambda+1)\delta_1\eta_2\eta_1$$

and

$$\rho_2 = (\alpha+\lambda+2)\delta_1 + \frac{\Gamma(\alpha+\lambda+3)}{\Gamma(\alpha+2)\Gamma(+2)}\delta_2\eta_1.$$

In a similar manner, the rest of the components can be obtained. Further, the approximate solution of $\widetilde{x}(t;r)$ and $\widetilde{y}(t;r)$ from Eq. (6.7) can be obtained.

In the special case, one may see that for $r = 1$, the fuzzy initial conditions convert to crisp initial conditions and the solutions obtained by the applied method exactly agree with the solutions of Das and Gupta (2011).

6.2 NUMERICAL RESULTS OF FUZZY ARBITRARY-ORDER PREDATOR–PREY EQUATIONS

In this section, numerical solution of fuzzy arbitrary-order predator–prey equations using HPM has been presented. It is a gigantic task to include here all the results with respect to various parameters involved in the corresponding equation. As such, few representative results are reported.

Fuzzy fractional predator–prey equation for $a(t) = t$, $b(t) = 1$, $c(t) = 1$, and $d(t) = t$ with fuzzy initial condition as discussed earlier is solved by HPM. Numerical results are depicted in Tables 6.1–6.4, respectively, by varying t from 0 to 0.8 and keeping r as constant for different types of fractional order, namely $\alpha = \lambda = 1/3$, $\alpha = \lambda = 1/2$, $\alpha = \lambda = 2/3$ and for integer order $\alpha = \lambda = 1$. Crisp results are also included in these tables. Also, from Tables 6.1–6.4, it can be concluded that the crisp solution lies in between the lower and upper bounds of the fuzzy and interval solution.

Next, by varying both t and r from 0 to 0.6 and 0 to 1 for $\alpha = \lambda = 1/3$, $\alpha = \lambda = 1/2$, $\alpha = \lambda = 2/3$, and $\alpha = \lambda = 1$, results are depicted, respectively, in Figs. 6.1–6.4.

TABLE 6.1　Fuzzy Solution for $\alpha = \lambda = 1/3$

$t \rightarrow$		0	0.4	0.6	0.8
$r = 0$	$[\underline{x}(t;r),\ \overline{x}(t;r)]$	[1.2, 1.4]	[−0.5433, 1.4637]	[−1.2409, 1.8615]	[−2.0396, 2.4442]
	$[\underline{y}(t;r),\ \overline{y}(t;r)]$	[0.5, 0.7]	[0.2490, 2.2562]	[−0.3048, 2.7977]	[−1.0008, 3.4830]
$r = 0.4$	$[\underline{x}(t;r),\ \overline{x}(t;r)]$	[1.24, 1.36]	[−0.0701, 1.1343]	[−0.5049, 1.3569]	[−0.9748, 1.7161]
	$[\underline{y}(t;r),\ \overline{y}(t;r)]$	[0.54, 0.66]	[0.5893, 1.7938]	[0.2243, 2.0862]	[−0.2283, 2.4626]
$r = 0.8$	$[\underline{x}(t;r),\ \overline{x}(t;r)]$	[1.28, 1.32]	[0.3673, 0.7688]	[0.1735, 0.7942]	[0.0063, 0.9034]
	$[\underline{y}(t;r),\ \overline{y}(t;r)]$	[0.58, 0.62]	[0.9603, 1.3618]	[0.7994, 1.4201]	[0.6065, 1.5036]
$r = 1$	$[\underline{x}(t;r),\ \overline{x}(t;r)]$	[1.3, 1.3]	[0.5725, 0.5725]	[0.4911, 0.4911]	[0.4654, 0.4654]
	$[\underline{y}(t;r),\ \overline{y}(t;r)]$	[0.6, 0.6]	[1.1572, 1.1572]	[1.1040, 1.1040]	[1.0473, 1.0473]
Das and Gupta (2011)	$x(t)$	1.3	0.5725	0.4911	0.4654
	$y(t)$	0.6	1.1572	1.1040	1.0473

TABLE 6.2　Fuzzy Solution for $\alpha = \lambda = 1/2$

$t \rightarrow$		0	0.4	0.6	0.8
$r = 0$	$[\underline{x}(t;r),\ \overline{x}(t;r)]$	[1.2, 1.4]	[0.0324, 1.3106]	[−0.5382, 1.5688]	[−1.2463, 1.9901]
	$[\underline{y}(t;r),\ \overline{y}(t;r)]$	[0.5, 0.7]	[0.5133, 1.7916]	[0.1336, 2.2407]	[−0.4293, 2.8071]
$r = 0.4$	$[\underline{x}(t;r),\ \overline{x}(t;r)]$	[1.24, 1.36]	[0.3276, 1.0946]	[−0.0441, 1.2202]	[−0.4821, 1.4601]
	$[\underline{y}(t;r),\ \overline{y}(t;r)]$	[0.54, 0.66]	[0.7347, 1.5017]	[0.4964, 1.7609]	[0.1296, 2.0719]
$r = 0.8$	$[\underline{x}(t;r),\ \overline{x}(t;r)]$	[1.28, 1.32]	[0.6030, 0.8587]	[0.4136, 0.8352]	[0.2238, 0.8713]
	$[\underline{y}(t;r),\ \overline{y}(t;r)]$	[0.58, 0.62]	[0.9732, 1.2289]	[0.8886, 1.3101]	[0.7330, 1.3805]
$r = 1$	$[\underline{x}(t;r),\ \overline{x}(t;r)]$	[1.3, 1.3]	[0.7333, 0.7333]	[0.6289, 0.6289]	[0.5548, 0.5548]
	$[\underline{y}(t;r),\ \overline{y}(t;r)]$	[0.6, 0.6]	[1.0989, 1.0989]	[1.0957, 1.0957]	[1.0512, 1.0512]
Das and Gupta (2011)	$x(t)$	1.3	0.8832	0.7333	0.6289
	$y(t)$	0.6	1.0117	1.0989	1.0957

TABLE 6.3   Fuzzy Solution $\alpha = \lambda = 2/3$

$t \rightarrow$		0	0.4	0.6	0.8
$r = 0$	$[\underline{x}(t;r),\ \overline{x}(t;r)]$	[1.2, 1.4]	[0.4063, 1.2525]	[−0.0247, 1.3982]	[−0.5857, 1.6804]
	$[\underline{y}(t;r),\ \overline{y}(t;r)]$	[0.5, 0.7]	[0.6337, 1.4798]	[0.4174, 1.8404]	[0.0216, 2.2878]
$r = 0.4$	$[\underline{x}(t;r),\ \overline{x}(t;r)]$	[1.24, 1.36]	[0.5967, 1.1044]	[0.3030, 1.1569]	[−0.0563, 1.3035]
	$[\underline{y}(t;r),\ \overline{y}(t;r)]$	[0.54, 0.66]	[0.7842, 1.2919]	[0.6663, 1.5202]	[0.4160, 1.776]
$r = 0.8$	$[\underline{x}(t;r),\ \overline{x}(t;r)]$	[1.28, 1.32]	[0.7765, 0.9457]	[0.6093, 0.8940]	[0.4350, 0.8883]
	$[\underline{y}(t;r),\ \overline{y}(t;r)]$	[0.58, 0.62]	[0.9440, 1.1133]	[0.9330, 1.2177]	[0.8399, 1.2933]
$r = 1$	$[\underline{x}(t;r),\ \overline{x}(t;r)]$	[1.3, 1.3]	[0.8624, 0.8624]	[0.75442, 0.75442]	[0.6664, 0.6664]
	$[\underline{y}(t;r),\ \overline{y}(t;r)]$	[0.6, 0.6]	[1.0275, 1.0275]	[1.0731, 1.0731]	[1.0629, 1.0629]
Das and Gupta (2011)	$x(t)$	1.3	0.8624	0.75442	0.6664
	$y(t)$	0.6	1.0275	1.0731	1.0629

TABLE 6.4   Fuzzy Solution for $\alpha = \lambda = 1$

$t \rightarrow$		0	0.4	0.6	0.8
$r = 0$	$[\underline{x}(t;r),\ \overline{x}(t;r)]$	[1.2, 1.4]	[0.8007, 1.2612]	[0.5579, 1.2818]	[0.2426, 1.3830]
	$[\underline{y}(t;r),\ \overline{y}(t;r)]$	[0.5, 0.7]	[0.6634, 1.1239]	[0.6425, 1.3665]	[0.5114, 1.6518]
$r = 0.4$	$[\underline{x}(t;r),\ \overline{x}(t;r)]$	[1.24, 1.36]	[0.8992, 1.1755]	[0.7175, 1.1518]	[0.5000, 1.1843]
	$[\underline{y}(t;r),\ \overline{y}(t;r)]$	[0.54, 0.66]	[0.7495, 1.0258]	[0.7744, 1.2088]	[0.7155, 1.3998]
$r = 0.8$	$[\underline{x}(t;r),\ \overline{x}(t;r)]$	[1.28, 1.32]	[0.9945, 1.0866]	[0.8697, 1.0145]	[0.7428, 0.9709]
	$[\underline{y}(t;r),\ \overline{y}(t;r)]$	[0.58, 0.62]	[0.8386, 0.9307]	[0.9128, 1.0576]	[0.9316, 1.1597]
$r = 1$	$[\underline{x}(t;r),\ \overline{x}(t;r)]$	[1.3, 1.3]	[1.0409, 1.0409]	[0.9430, 0.9430]	[0.8587, 0.8587]
	$[\underline{y}(t;r),\ \overline{y}(t;r)]$	[0.6, 0.6]	[0.8843, 0.8843]	[0.9844, 0.9844]	[1.0442, 1.0442]
Das and Gupta (2011)	$x(t)$	1.3	1.0409	0.9430	0.8587
	$y(t)$	0.6	0.8843	0.9844	1.0442

Now for $r = 0.6$ and 1, the interval solution with various order derivatives, namely $\alpha = \lambda = 1/3$, $\alpha = \lambda = 1/2$, $\alpha = \lambda = 2/3$, and $\alpha = \beta = 1$ are computed and those are shown in Figs. 6.5–6.8. It is interesting to note that for both the cases, the lower and upper bounds of the fuzzy solutions are the same for $r = 1$, which approximately matches with the crisp solution of Das and Gupta (2011).

From Figs. 6.5–6.8, one may see that prey decreases and predator increases with time t. One may note that it takes more time for meeting predator–prey populations as the fractional time derivative increases and finally takes the maximum time for $\alpha = \lambda = 1$.

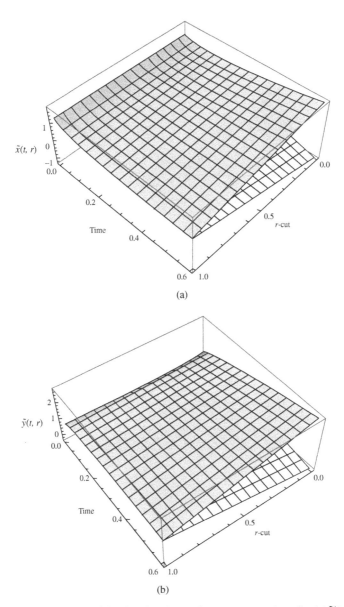

Figure 6.1 Fuzzy solution of fractional-order predator–prey equations for (a) $\widetilde{x}(t; r)$ and (b) $\widetilde{y}(t; r)$ where $t \in [0, 0.6]$ and $r \in [0, 1]$ when $\alpha = \lambda = 1/3$

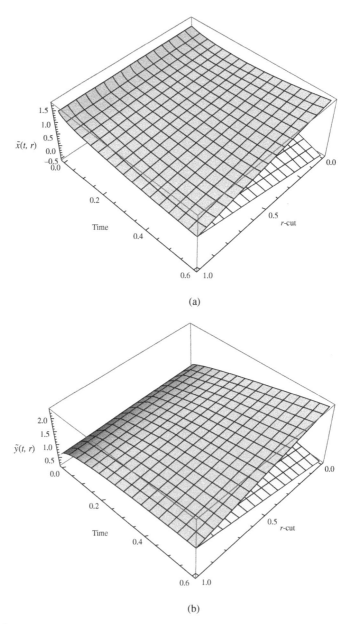

(a)

(b)

Figure 6.2   Fuzzy solution of fractional-order predator–prey equations for (a) $\widetilde{x}(t; r)$ and (b) $\widetilde{y}(t; r)$ where $t \in [0, 0.6]$ and $r \in [0, 1]$ when $\alpha = \lambda = 1/2$

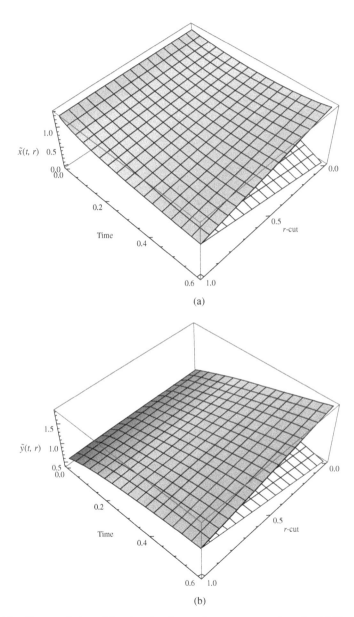

Figure 6.3 Fuzzy solution of fractional-order predator–prey equations for (a) $\widetilde{x}(t;r)$ and (b) $\widetilde{y}(t;r)$ where $t \in [0, 0.6]$ and $r \in [0, 1]$ when $\alpha = \lambda = 2/3$

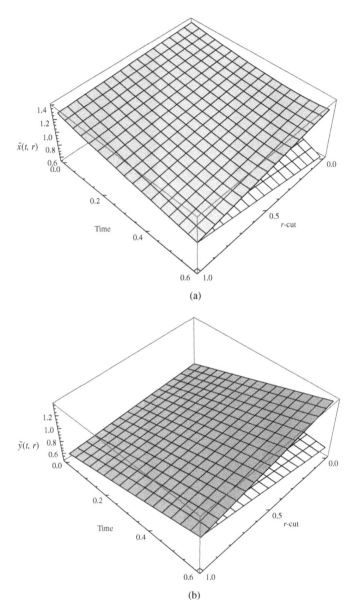

(a)

(b)

Figure 6.4 Fuzzy solution of fractional-order predator–prey equations for (a) $\widetilde{x}(t; r)$ and (b) $\widetilde{y}(t; r)$ where $t \in [0, 0.6]$ and $r \in [0, 1]$ when $\alpha = \lambda = 1$

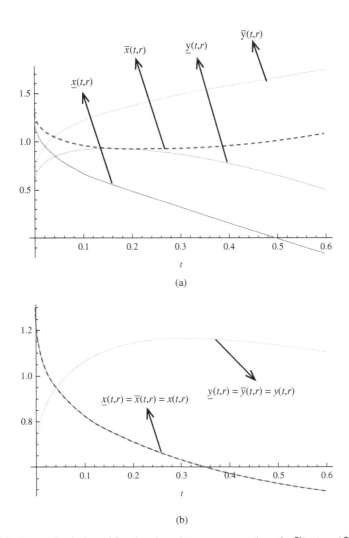

Figure 6.5 Interval solution of fractional predator–prey equations for $\widetilde{x}(t; r)$ and $\widetilde{y}(t; r)$ with $\alpha = \lambda = 1/3$ at (a) $r = 0.6$ and (b) $r = 1$

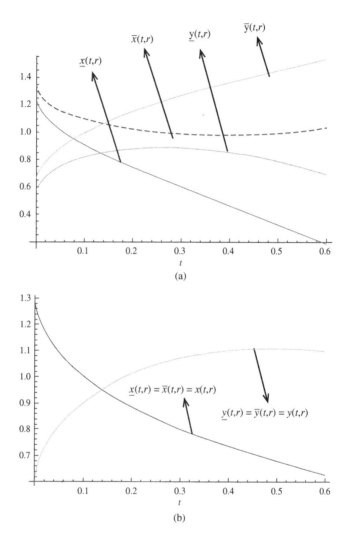

Figure 6.6 Interval solution of fractional predator–prey equations for $\widetilde{x}(t; r)$ and $\widetilde{y}(t; r)$ with $\alpha = \lambda = 1/2$ at (a) $r = 0.6$ and (b) $r = 1$

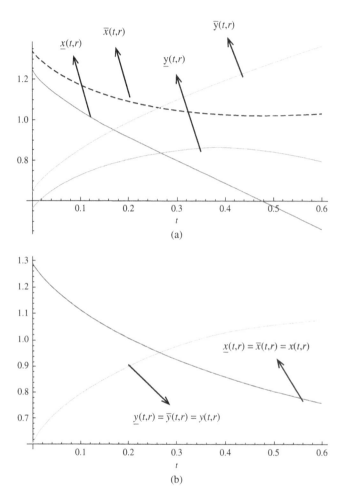

Figure 6.7 Interval solution of fractional predator–prey equations for $\widetilde{x}(t)$ and $\widetilde{y}(t)$ with $\alpha = \lambda = 2/3$ at (a) $r = 0.6$ and (b) $r = 1$

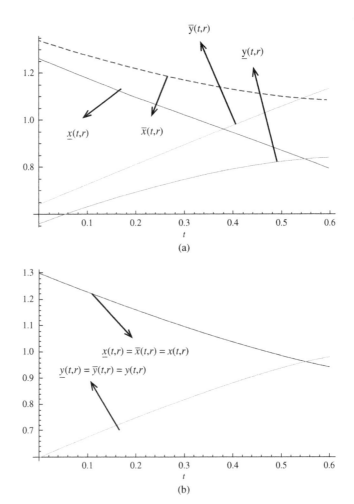

Figure 6.8 Interval solution of fractional predator–prey equations for $\widetilde{x}(t; r)$ and $\widetilde{y}(t; r)$ with $\alpha = \lambda = 1$ at (a) $r = 0.6$ and (b) $r = 1$

BIBLIOGRAPHY

Agrawal RP, Lakshmikantham V, Nieto JJ. On the concept of solution for fractional differential equations with uncertainty. *Nonlinear Anal* 2010;**72**:2859–2862.

Ahmed E, El-Sayed AMA, El-Saka HAA. Equilibrium points, stability and numerical solutions of fractional-order predator–prey and rabies models. *J Math Anal Appl* 2007;**325**:542–553.

Chakraverty S. *Mathematics of Uncertainty Modeling in the Analysis of Engineering and Science Problems*. USA: IGI Global Publication; 2014.

Chalco-Cano Y, Roman-Flores H. On new solutions of fuzzy differential equations. *Chaos Solitons Fract* 2008;**38**:112–119.

Das S, Gupta PK. A mathematical model on fractional Lotka–Volterra equations. *J Theor Biol* 2011;**277**:1–6.

El-Sayed AMA, Rida SZ, Arafa AAM. Exact solutions of fractional-order biological population model. *Commun Theor Phys* 2009;**52**:992–996.

Hanss M. *Applied Fuzzy Arithmetic: An Introduction with Engineering Applications*. Berlin: Springer-Verlag; 2005.

He JH. Homotopy Perturbation Technique. *Comput Methods Appl Mech Eng* 1999;**178**:257–262.

He JH. A coupling method of a homotopy technique and a perturbation technique for non-linear problems. *Int J Nonlinear Mech* 2000;**35**:37–43.

Jaulin L, Kieffer M, Didri OT, Walter E. *Applied Interval Analysis*. Springer; 2001.

Liu Y, Xin B. Numerical solutions of a fractional predator–prey system. *Adv Differ Eqs* 2011;**2011**:1–11.

Lotka AJ. *Elements of Physical Biology*. Baltimore Williams & Wilkins Company; 1925.

Moore RE. *Interval Analysis*. Englewood Cliffs: Prentice Hall; 1966.

Podlubny I. *Fractional Differential Equations*. New York, NY: Academic Press; 1999.

Ross TJ. *Fuzzy Logic with Engineering Applications*. New York: John Wiley & Sons; 2004.

Shakeri F, Dehghan M. Numerical solution of a biological population model using He's variational iteration method. *Comput Math Appl* 2007;**54**:1197–1209.

Tapaswini S, Chakraverty S. Numerical solution of fuzzy arbitrary order predator–prey equations. *Appl Appl Math Int J* 2013;**8**:647–672.

Zimmermann HJ. *Fuzzy Set Theory and Its Application*. London: Kluwer Academic Publishers; 2001.

7

FUZZY FRACTIONAL CHEMICAL PROBLEMS

The target of this chapter is to investigate the numerical solution of uncertain arbitrary-order Rossler's system using homotopy perturbation method (HPM). Rossler's system was found to be useful in the modeling of equilibrium in chemical reactions. Here also, uncertainty in the initial conditions is handled by triangular convex, normalized fuzzy sets. Numerical solution of arbitrary-order Rossler's system with crisp initial condition is also included.

7.1 ARBITRARY-ORDER ROSSLER'S SYSTEMS

Otto Rossler proposed Rossler's system with strange attractor in 1976, but the original theoretical equations were later found to be useful in modeling equilibrium in chemical reactions. This attractor has only one manifold and can be obtained as a solution of the following equations (Petras, 2011):

$$\frac{dx(t)}{dt} = -(y(t) + z(t))$$

$$\frac{dy(t)}{dt} = x(t) + ay(t) \tag{7.1}$$

$$\frac{dz(t)}{dt} = b + z(t)(x(t) - c),$$

where x, y, z are the state variables, and a, b, c are three system parameters.

Fuzzy Arbitrary Order System: Fuzzy Fractional Differential Equations and Applications, First Edition.
Snehashish Chakraverty, Smita Tapaswini, and Diptiranjan Behera.
© 2016 John Wiley & Sons, Inc. Published 2016 by John Wiley & Sons, Inc.

Here, we have considered a fractional-order system of the Rossler's system as follows:

$$D_t^\alpha x(t) = -(y(t) + z(t))$$
$$D_t^\beta y(t) = x(t) + ay(t) \tag{7.2}$$
$$D_t^\gamma z(t) = b + z(t)(x(t) - c),$$

where $0 < \alpha, \beta, \gamma \le 1$ is derivative order and $a, b,$ and c are functions of time t. Related fuzzy Rossler's system may now be defined as

$$D_t^\alpha \widetilde{x}(t) = -(\widetilde{y}(t) + \widetilde{z}(t))$$
$$D_t^\beta \widetilde{y}(t) = \widetilde{x}(t) + a(t)\widetilde{y}(t) \tag{7.3}$$
$$D_t^\gamma \widetilde{z}(t) = b(t) + \widetilde{z}(t)(\widetilde{x}(t) - c(t)).$$

7.2   HPM SOLUTION OF UNCERTAIN ARBITRARY-ORDER ROSSLER'S SYSTEM

Let us consider the aforementioned fuzzy fractional Rossler's system as

$$D_t^\alpha \widetilde{x}(t) + \widetilde{y}(t) + \widetilde{z}(t) = 0$$
$$D_t^\beta \widetilde{y}(t) - \widetilde{x}(t) - a(t)\widetilde{y}(t) = 0 \tag{7.4}$$
$$D_t^\gamma \widetilde{z}(t) - b(t) - \widetilde{z}(t)(\widetilde{x}(t) - c(t)) = 0$$

with initial conditions in terms of triangular fuzzy number, namely

$$\widetilde{x}(0) = (a_1, b_1, c_1), \widetilde{y}(0) = (a_2, b_2, c_2), \text{ and } \widetilde{z}(0) = (a_3, b_3, c_3).$$

Using r-cut, triangular fuzzy initial conditions become

$$\widetilde{x}(0) = [r(b_1 - a_1) + a_1, c_1 - r(c_1 - b_1)],$$
$$\widetilde{y}(0) = [r(b_2 - a_2) + a_2, c_2 - r(c_2 - b_2)], \text{ and}$$
$$\widetilde{y}(0) = [r(b_2 - a_2) + a_2, c_2 - r(c_2 - b_2)], \quad 0 \le r \le 1.$$

Here, $\widetilde{x}(t) = [\underline{x}(t), \overline{x}(t)]$, $\widetilde{y}(t) = [\underline{y}(t), \overline{y}(t)]$, and $\widetilde{z}(t) = [\underline{z}(t), \overline{z}(t)]$ are fuzzy functions of t.

Using Hukahara differentiability, the fuzzy fractional Rossler's system, that is, Eq. (7.4) may be reduced to a set of ordinary differential equations

$$D_t^\alpha \underline{x}(t; r) + \underline{y}(t; r) + \underline{z}(t; r) = 0$$
$$D_t^\alpha \overline{x}(t; r) + \overline{y}(t; r) + \overline{z}(t; r) = 0$$

$$D_t^\beta \underline{y}(t; r) - \bar{x}(t; r) - a(t)\bar{y}(t; r) = 0 \tag{7.5}$$

$$D_t^\beta \bar{y}(t; r) - \underline{x}(t; r) - a(t)\underline{y}(t; r) = 0$$

$$D_t^\gamma \underline{z}(t; r) - b(t) - \bar{z}(t; r)\bar{x}(t; r) + c(t)\underline{z}(t; r) = 0$$

$$D_t^\gamma \bar{z}(t; r) - b(t) - \underline{z}(t; r)\underline{x}(t; r) + c(t)\bar{z}(t; r) = 0.$$

We can readily construct a homotopy for every equation of (7.5), which satisfies

$$(1 - p)D_t^\alpha \underline{x}(t; r) + p\left[D_t^\alpha \underline{x}(t; r) + \underline{y}(t; r) + \underline{z}(t; r)\right] = 0$$

$$(1 - p)D_t^\alpha \bar{x}(t; r) + p\left[D_t^\alpha \bar{x}(t; r) + \bar{y}(t; r) + \bar{z}(t; r)\right] = 0$$

$$(1 - p)D_t^\beta \underline{y}(t; r) + p\left[D_t^\beta \underline{y}(t; r) - \bar{x}(t; r) - a(t)\bar{y}(t; r)\right] = 0$$

$$(1 - p)D_t^\beta \bar{y}(t; r) + p\left[D_t^\beta \bar{y}(t; r) - \underline{x}(t; r) - a(t)\underline{y}(t; r)\right] = 0 \tag{7.6}$$

$$(1 - p)D_t^\gamma \underline{z}(t; r) + p\left[D_t^\gamma \underline{z}(t; r) - b(t) - \bar{z}(t; r)\bar{x}(t; r) + c(t)\underline{z}(t; r)\right] = 0$$

$$(1 - p)D_t^\gamma \underline{z}(t; r) + p\left[D_t^\gamma \bar{z}(t; r) - b(t) - \underline{z}(t; r)\underline{x}(t; r) + c(t)\bar{z}(t; r)\right] = 0.$$

One may try to obtain a solution of Eq. (7.4) in the form

$$\underline{x}(t; r) = \underline{x}_0(t; r) + p\underline{x}_1(t; r) + p^2\underline{x}_2(t; r) + \cdots, \tag{7.7}$$

$$\bar{x}(t; r) = \bar{x}_0(t; r) + p\bar{x}_1(t; r) + p^2\bar{x}_2(t; r) + \cdots, \tag{7.8}$$

$$\underline{y}(t; r) = \underline{y}_0(t; r) + p\underline{y}_1(t; r) + p^2\underline{y}_2(t; r) + \cdots, \tag{7.9}$$

$$\bar{y}(t; r) = \bar{y}_0(t; r) + p\bar{y}_1(t; r) + p^2\bar{y}_2(t; r) + \cdots, \tag{7.10}$$

$$\underline{z}(t; r) = \underline{z}_0(t; r) + p\underline{z}_1(t; r) + p^2\underline{z}_2(t; r) + \cdots, \tag{7.11}$$

$$\bar{z}(t; r) = \bar{z}_0(t; r) + p\bar{z}_1(t; r) + p^2\bar{z}_2(t; r) + \cdots, \tag{7.12}$$

where $\underline{x}_i(t; r), \bar{x}_i(t; r), \underline{y}_i(t; r), \bar{y}_i(t; r), \underline{z}_i(t),$ and $\bar{z}_i(t), i = 0, 1, 2, \ldots$ are functions yet to be determined. Substituting Eqs. (7.7)–(7.12) into Eqs. (7.6), respectively, and equating the terms with the identical powers of p, we have

$$p^0 : \begin{cases} D_t^\alpha \underline{x}_0(t; r) = 0, \\ D_t^\alpha \bar{x}_0(t; r) = 0, \\ D_t^\beta \underline{y}_0(t; r) = 0, \\ D_t^\beta \bar{y}_0(t; r) = 0, \\ D_t^\gamma \underline{z}_0(t; r) = 0, \\ D_t^\gamma \bar{z}_0(t; r) = 0, \end{cases}$$

$$p^1 : \begin{cases} D_t^\alpha \underline{x}_1(t;r) = -\underline{y}_0(t;r) - \underline{z}_0(t;r), \\ D_t^\alpha \overline{x}_1(t;r) = -\overline{y}_0(t;r) - \overline{z}_0(t;r), \\ D_t^\beta \underline{y}_1(t;r) = \overline{x}_0(t) + a(t)\overline{y}_0(t;r), \\ D_t^\beta \overline{y}_1(t;r) = \underline{x}_0(t;r) + a(t)\underline{y}_0(t;r), \\ D_t^\gamma \underline{z}_1(t;r) = b(t) + \overline{z}_0(t;r)\overline{x}_0(t;r) - c(t)\underline{z}_0(t;r), \\ D_t^\gamma \overline{z}_1(t;r) = b(t) + \underline{z}_0(t;r)\underline{x}_0(t;r) - c(t)\overline{z}_0(t;r), \end{cases} \qquad (7.13)$$

$$p^2 : \begin{cases} D_t^\alpha \underline{x}_2(t;r) = -\underline{y}_1(t;r) - \underline{z}_1(t;r), \\ D_t^\alpha \overline{x}_2(t;r) = -\overline{y}_1(t;r) - \overline{z}_1(t;r), \\ D_t^\beta \underline{y}_2(t;r) = \overline{x}_1(t;r) + a(t)\overline{y}_1(t;r), \\ D_t^\beta \overline{y}_2(t;r) = \underline{x}_1(t;r) + a(t)\underline{y}_1(t;r), \\ D_t^\gamma \underline{z}_2(t;r) = \overline{z}_1(t)\overline{x}_0(t) + \overline{z}_0(t)\overline{x}_1(t) - c(t)\underline{z}_1(t), \\ D_t^\gamma \overline{z}_2(t;r) = \underline{z}_1(t;r)\underline{x}_0(t;r) + \underline{z}_0(t;r)\underline{x}_1(t;r) - c(t)\overline{z}_1(t;r), \end{cases}$$

$$p^3 : \begin{cases} D_t^\alpha \underline{x}_3(t;r) = -\underline{y}_2(t;r) - \underline{z}_2(t;r), \\ D_t^\alpha \overline{x}_3(t;r) = -\overline{y}_2(t;r) - \overline{z}_2(t;r), \\ D_t^\beta \underline{y}_3(t;r) = \overline{x}_2(t;r) + a(t)\overline{y}_2(t;r), \\ D_t^\beta \overline{y}_3(t;r) = \underline{x}_2(t;r) + a(t)\underline{y}_2(t;r), \\ D_t^\gamma \underline{z}_3(t;r) = \overline{z}_0(t;r)\overline{x}_2(t;r) + \overline{z}_1(t;r)\overline{x}_1(t;r) + \overline{z}_2(t;r)\overline{x}_0(t;r) + \underline{z}_2(t;r), \\ D_t^\gamma \overline{z}_3(t;r) = \underline{z}_0(t;r)\underline{x}_2(t;r) + \underline{z}_1(t;r)\underline{x}_1(t;r) + \underline{z}_2(t;r)\underline{x}_0(t;r) + \overline{z}_2(t;r), \end{cases}$$

and so on.

The method is based on applying the operators J_t^α, J_t^β, and J_t^γ (the inverse operators of Caputo derivatives D_t^α, D_t^β, and D_t^γ, respectively) on both sides of Eq. (7.13), then we can calculate the approximate solution bounds $\underline{x}(t;r) = \sum_{n=0}^\infty \underline{x}_n(t;r)$, $\underline{y}(t;r) = \sum_{n=0}^\infty \underline{y}_n(t;r)$, $\underline{z}(t;r) = \sum_{n=0}^\infty \underline{z}_n(t;r)$ and $\overline{x}(t;r) = \sum_{n=0}^\infty \overline{x}_n(t;r)$, $\overline{y}(t;r) = \sum_{n=0}^\infty \overline{y}_n(t;r)$, $\overline{z}(t;r) = \sum_{n=0}^\infty \overline{z}_n(t;r)$, term by term. Otherwise, an approximation to the solutions would be achieved by computing some terms, say k, as

$$\underline{x}(t;r) = \lim_{k\to\infty} \underline{\xi}_k(t;r), \; \underline{y}(t;r) = \lim_{k\to\infty} \underline{\phi}_k(t;r) \text{ and } \underline{z}(t;r) = \lim_{k\to\infty} \underline{\psi}_k(t;r) \qquad (7.14)$$

$$\overline{x}(t;r) = \lim_{k\to\infty} \overline{\xi}_k(t;r), \; \overline{y}(t;r) = \lim_{k\to\infty} \overline{\phi}_k(t;r) \text{ and } \overline{z}(t;r) = \lim_{k\to\infty} \overline{\psi}_k(t;r)$$

where, $\underline{\xi}_k(t;r) = \sum_{n=0}^k \underline{x}_n(t;r)$, $\underline{\phi}_k(t;r) = \sum_{n=0}^k \underline{y}_n(t;r)$, $\underline{\psi}_k(t;r) = \sum_{n=0}^k \underline{z}_n(t;r)$, $\overline{\xi}_k(t;r) = \sum_{n=0}^k \overline{x}_n(t;r)$, $\overline{\phi}_k(t;r) = \sum_{n=0}^k \overline{y}_n(t;r)$, and $\overline{\psi}_k(t;r) = \sum_{n=0}^k \overline{z}_n(t;r)$.

7.3 PARTICULAR CASE

Let us consider $a(t) = t, b(t) = t, c(t) = 1$ and initial conditions in terms of triangular fuzzy numbers, namely $x(0) = (0.4, 0.5, 0.6)$, $y(0) = (1.4, 1.5, 1.6)$, and $z(0) = (0, 0.1, 0.2)$.

Through r-cut approach, initial condition becomes $x(0) = [0.1r + 0.4, 0.6 - 0.1r]$ $= [\delta_1, \delta_2]$, $y(0) = [0.1r + 1.4, 1.6 - 0.1r] = [\eta_1, \eta_2]$, and $z(0) = [0.1r, 0.2 - 0.1r] = [\mu_1, \mu_2]$. Solving Eq. (7.13), one may have

$$\underline{x}_0(t; r) = \delta_1,$$

$$\overline{x}_0(t; r) = \delta_2,$$

$$\underline{y}_0(t; r) = \eta_1,$$

$$\overline{y}_0(t; r) = \eta_2,$$

$$\underline{z}_0(t; r) = \mu_1,$$

$$\overline{z}_0(t; r) = \mu_2,$$

$$\underline{x}_1(t; r) = -(\eta_1 + \mu_1)\frac{t^\alpha}{\Gamma(\alpha + 1)},$$

$$\overline{x}_1(t; r) = -(\eta_2 + \mu_2)\frac{t^\alpha}{\Gamma(\alpha + 1)},$$

$$\underline{y}_1(t; r) = \delta_2\frac{t^\beta}{\Gamma(\beta + 1)} + \eta_2\frac{t^{\beta+1}}{\Gamma(\beta + 2)},$$

$$\overline{y}_1(t; r) = \delta_1\frac{t^\beta}{\Gamma(\beta + 1)} + \eta_1\frac{t^{\beta+1}}{\Gamma(\beta + 2)},$$

$$\underline{z}_1(t; r) = (\mu_2\delta_2 - \mu_1)\frac{t^\gamma}{\Gamma(\gamma + 1)} + \frac{t^{\gamma+1}}{\Gamma(\gamma + 2)},$$

$$\overline{z}_1(t; r) = (\mu_1\delta_1 - \mu_2)\frac{t^\gamma}{\Gamma(\gamma + 1)} + \frac{t^{\gamma+1}}{\Gamma(\gamma + 2)},$$

$$\underline{x}_2(t; r) = -\delta_2\frac{t^{\alpha+\beta}}{\Gamma(\alpha + \beta + 1)} - \eta_2\frac{t^{\alpha+\beta+1}}{\Gamma(\alpha + \beta + 2)} - \frac{t^{\alpha+\gamma+1}}{\Gamma(\alpha + \gamma + 2)}$$
$$- (\mu_2\delta_2 - \mu_1)\frac{t^{\alpha+\gamma}}{\Gamma(\alpha + \gamma + 1)},$$

$$\overline{x}_2(t; r) = -\delta_1\frac{t^{\alpha+\beta}}{\Gamma(\alpha + \beta + 1)} - \eta_1\frac{t^{\alpha+\beta+1}}{\Gamma(\alpha + \beta + 2)} - \frac{t^{\alpha+\gamma+1}}{\Gamma(\alpha + \gamma + 2)}$$
$$- (\mu_1\delta_1 - \mu_2)\frac{t^{\alpha+\gamma}}{\Gamma(\alpha + \gamma + 1)},$$

$$\underline{y}_2(t; r) = -(\eta_2 + \mu_2)\frac{t^{\alpha+\beta}}{\Gamma(\alpha + \beta + 1)} + \delta_1\,(\beta + 1)\frac{t^{2\beta+1}}{\Gamma(2\beta + 2)}$$

$$+ \eta_1(\beta + 2)\frac{t^{2\beta+2}}{\Gamma(2\beta + 3)},$$

$$\overline{y}_2(t; r) = -(\eta_1 + \mu_1)\frac{t^{\alpha+\beta}}{\Gamma(\alpha + \beta + 1)} + \delta_2\,(\beta + 1)\frac{t^{2\beta+1}}{\Gamma(2\beta + 2)} + \eta_2(\beta + 2)\frac{t^{2\beta+2}}{\Gamma(2\beta + 3)},$$

$$\underline{z}_2(t; r) = (\delta_2 - 1)\frac{t^{2\gamma}}{\Gamma(2\gamma + 2)} + (\delta_2\mu_1\delta_1 - 2\mu_2\delta_2 + \mu_1)\frac{t^{2\gamma}}{\Gamma(2\gamma + 1)}$$

$$- \mu_2(\eta_2 + \mu_2)\frac{t^{\alpha+\gamma}}{\Gamma(\alpha + \gamma + 1)},$$

$$\overline{z}_2(t; r) = (\delta_1 - 1)\frac{t^{2\gamma}}{\Gamma(2\gamma + 2)} + (\delta_1\mu_2\delta_2 - 2\mu_1\delta_1 + \mu_2)\frac{t^{2\gamma}}{\Gamma(2\gamma + 1)}$$

$$- \mu_1(\eta_1 + \mu_1)\frac{t^{\alpha+\gamma}}{\Gamma(\alpha + \gamma + 1)},$$

$$\underline{x}_3(t; r) = (\eta_2 + \mu_2)\frac{t^{2\alpha+\beta}}{\Gamma(2\alpha + \beta + 1)} - \delta_1(\beta + 1)\frac{t^{\alpha+2\beta+1}}{\Gamma(\alpha + 2\beta + 2)}$$

$$- \eta_1(\beta + 2)\frac{t^{\alpha+2\beta+2}}{\Gamma(\alpha + 2\beta + 3)} - (\delta_2 - 1)\frac{t^{\alpha+2\gamma+1}}{\Gamma(\alpha + 2\gamma + 2)}$$

$$- (\mu_1\delta_1\delta_2 - 2\mu_2\delta_2 + \mu_1)\frac{t^{\alpha+2\gamma}}{\Gamma(\alpha + 2\gamma + 1)} + \mu_2(\eta_2 + \mu_2)\frac{t^{2\alpha+\gamma}}{\Gamma(2\alpha + \gamma + 1)},$$

$$\overline{x}_3(t; r) = (\eta_1 + \mu_1)\frac{t^{2\alpha+\beta}}{\Gamma(2\alpha + \beta + 1)} - \delta_2(\beta + 1)\frac{t^{\alpha+2\beta+1}}{\Gamma(\alpha + 2\beta + 2)}$$

$$- \eta_2(\beta + 2)\frac{t^{\alpha+2\beta+2}}{\Gamma(\alpha + 2\beta + 3)} - (\delta_1 - 1)\frac{t^{\alpha+2\gamma+1}}{\Gamma(\alpha + 2\gamma + 2)}$$

$$- (\mu_2\delta_1\delta_2 - 2\mu_1\delta_1 + \mu_2)\frac{t^{\alpha+2\gamma}}{\Gamma(\alpha + 2\gamma + 1)} + \mu_1(\eta_1 + \mu_1)\frac{t^{2\alpha+\gamma}}{\Gamma(2\alpha + \gamma + 1)},$$

$$\underline{y}_3(t; r) = -\delta_1\frac{t^{\alpha+2\beta}}{\Gamma(\alpha + 2\beta + 1)} - \eta_1\frac{t^{\alpha+2\beta+1}}{\Gamma(\alpha + 2\beta + 2)} - \frac{t^{\alpha+\beta+\gamma+1}}{\Gamma(\alpha + \beta + \gamma + 2)}$$

$$- (\mu_1\delta_1 - \mu_2)\frac{t^{\alpha+\beta+\gamma}}{\Gamma(\alpha + \beta + \gamma + 1)} - (\eta_1 + \mu_1)(\alpha + \beta + 1)\frac{t^{\alpha+2\beta+1}}{\Gamma(\alpha + 2\beta + 2)}$$

$$+ \delta_2(\beta + 1)(2\beta + 2)\frac{t^{3\beta+2}}{\Gamma(3\beta + 3)} + \eta_2(\beta + 2)(2\beta + 3)\frac{t^{3\beta+3}}{\Gamma(3\beta + 4)},$$

$$\overline{y}_3(t; r) = -\delta_2\frac{t^{\alpha+2\beta}}{\Gamma(\alpha + 2\beta + 1)} - \eta_2\frac{t^{\alpha+2\beta+1}}{\Gamma(\alpha + 2\beta + 2)} - \frac{t^{\alpha+\beta+\gamma+1}}{\Gamma(\alpha + \beta + \gamma + 2)}$$

$$- (\mu_2\delta_2 - \mu_1)\frac{t^{\alpha+\beta+\gamma}}{\Gamma(\alpha + \beta + \gamma + 1)} - (\eta_2 + \mu_2)(\alpha + \beta + 1)\frac{t^{\alpha+2\beta+1}}{\Gamma(\alpha + 2\beta + 2)}$$

$$+ \delta_1(\beta + 1)(2\beta + 2)\frac{t^{3\beta+2}}{\Gamma(3\beta + 3)} + \eta_1(\beta + 2)(2\beta + 3)\frac{t^{3\beta+3}}{\Gamma(3\beta + 4)},$$

$$\underline{z}_3(t;r) = -\delta_1\mu_2\frac{t^{\alpha+\beta+\gamma}}{\Gamma(\alpha+\beta+\gamma+1)} - \eta_1\mu_2\frac{t^{\alpha+\beta+\gamma+1}}{\Gamma(\alpha+\beta+\gamma+2)}$$

$$+\left\{-\mu_2 - (\eta_2+\mu_2)\frac{\Gamma(\alpha+\gamma+2)}{\Gamma(\alpha+1)\Gamma(\gamma+2)}\right\}\frac{t^{\alpha+2\gamma+1}}{\Gamma(\alpha+2\gamma+2)}$$

$$+\left\{-\mu_1\mu_2\delta_1 - (\eta_2+\mu_2)\left(\mu_1\delta_1 - \mu_2\right)\frac{\Gamma(\alpha+\gamma+1)}{\Gamma(\alpha+1)\Gamma(\gamma+1)} - \mu_1\delta_2\eta_1\right.$$

$$\left.-\mu_1^2\delta_2 - \mu_2\eta_2\right\}\frac{t^{\alpha+2\gamma}}{\Gamma(\alpha+2\gamma+1)} + (\delta_1\delta_2 - 2\delta_2 + 1)\frac{t^{3\gamma+1}}{\Gamma(3\gamma+2)}$$

$$+\left(\mu_2\delta_1\delta_2^2 - 3\mu_1\delta_1\delta_2 + 3\mu_2\delta_2 - \mu_1\right)\frac{t^{3\gamma}}{\Gamma(3\gamma+1)},$$

$$\overline{z}_3(t;r) = -\delta_2\mu_1\frac{t^{\alpha+\beta+\gamma}}{\Gamma(\alpha+\beta+\gamma+1)} - \eta_2\mu_1\frac{t^{\alpha+\beta+\gamma+1}}{\Gamma(\alpha+\beta+\gamma+2)}$$

$$+\left\{-\mu_1 - (\eta_1+\mu_1)\frac{\Gamma(\alpha+\gamma+2)}{\Gamma(\alpha+1)\Gamma(\gamma+2)}\right\}\frac{t^{\alpha+2\gamma+1}}{\Gamma(\alpha+2\gamma+2)}$$

$$+\left\{-\mu_2\mu_1\delta_2 - (\eta_1+\mu_1)\left(\mu_2\delta_2 - \mu_1\right)\frac{\Gamma(\alpha+\gamma+1)}{\Gamma(\alpha+1)\Gamma(\gamma+1)} - \mu_2\delta_1\eta_2\right.$$

$$\left.-\mu_2^2\delta_1 - \mu_1\eta_1\right\}\frac{t^{\alpha+2\gamma}}{\Gamma(\alpha+2\gamma+1)} + (\delta_1\delta_2 - 2\delta_1 + 1)\frac{t^{3\gamma+1}}{\Gamma(3\gamma+2)}$$

$$+\left(\mu_1\delta_2\delta_1^2 - 3\mu_2\delta_1\delta_2 + 3\mu_1\delta_1 - \mu_2\right)\frac{t^{3\gamma}}{\Gamma(3\gamma+1)}.$$

In a similar manner, the rest of the components can be obtained. Further, one may get the approximate solution of $\tilde{x}(t)$, $\tilde{y}(t)$, and $\tilde{z}(t)$ from Eq. (7.14).

As discussed earlier, here fuzzy fractional Rossler's system with triangular fuzzy initial conditions has been considered. In the special case, one may see that for $r = 1$ in the r-cut form of fuzzy numbers, it converts to crisp initial conditions. So it is interesting to study fractional-order Rossler's system with crisp initial conditions as follows.

7.3.1 Special Case

Let us consider the crisp fractional-order Rossler's system as in Eq. (7.3) when $a(t) = t$, $b(t) = t$, $c(t) = 1$ with crisp initial condition $x(0) = \delta, y(0) = \eta$, and $y(0) = \mu$. Using HPM, one may have

$$x_0(t) = \delta,$$

$$y_0(t) = \eta,$$

$$z_0(t) = \mu,$$

$$x_1(t) = -(\eta + \mu)\frac{t^{\alpha}}{\Gamma(\alpha+1)},$$

$$y_1(t) = \delta \frac{t^\beta}{\Gamma(\beta + 1)} + \eta \frac{t^{\beta+1}}{\Gamma(\beta + 2)},$$

$$z_1(t) = \frac{t^{\gamma+1}}{\Gamma(\gamma + 2)} + \mu(\delta - 1)\frac{t^\gamma}{\Gamma(\gamma + 1)},$$

$$x_2(t) = -\delta \frac{t^{\alpha+\beta}}{\Gamma(\alpha + \beta + 1)} - \eta \frac{t^{\alpha+\beta+1}}{\Gamma(\alpha + \beta + 2)} - \frac{t^{\alpha+\gamma+1}}{\Gamma(\alpha + \gamma + 2)} - \mu(\delta - 1)\frac{t^{\alpha+\gamma}}{\Gamma(\alpha + \gamma + 1)},$$

$$y_2(t) = -(\eta + \mu)\frac{t^{\alpha+\beta}}{\Gamma(\alpha + \beta + 1)} + \delta(\beta + 1)\frac{t^{2\beta+1}}{\Gamma(2\beta + 2)} + \eta\,(\beta + 2)\frac{t^{2\beta+2}}{\Gamma(2\beta + 3)},$$

$$z_2(t) = -\mu(\eta + \mu)\frac{t^{\alpha+\gamma}}{\Gamma(\alpha + \gamma + 1)} + (\delta - 1)\frac{t^{2\gamma+1}}{\Gamma(2\gamma + 2)} + \mu(\delta - 1)^2\frac{t^{2\gamma}}{\Gamma(2\gamma + 1)},$$

$$x_3(t) = (\eta + \mu)\frac{t^{2\alpha+\beta}}{\Gamma(2\alpha + \beta + 1)} - \delta(\beta + 1)\frac{t^{\alpha+2*\beta+1}}{\Gamma(\alpha + 2\beta + 2)} - \eta(\beta + 2)\frac{t^{\alpha+2\beta+2}}{\Gamma(\alpha + 2\beta + 3)}$$
$$+ \mu(\eta + \mu)\frac{t^{2\alpha+\gamma}}{\Gamma(2\alpha + \gamma + 1)} - (\delta - 1)\frac{t^{\alpha+2\gamma+1}}{\Gamma(\alpha + 2\gamma + 2)}$$
$$- \mu(\delta - 1)^2 \frac{t^{\alpha+2\gamma}}{\Gamma(\alpha + 2\gamma + 1)},$$

$$y_3(t) = -\delta\frac{t^{\alpha+2\beta}}{\Gamma(\alpha + 2\beta + 1)} - \eta\frac{t^{\alpha+2\beta+1}}{\Gamma(\alpha + 2\beta + 2)} - \frac{t^{\alpha+\beta+\gamma+1}}{\Gamma(\alpha + \beta + \gamma + 2)}$$
$$- \mu(\delta - 1)\frac{t^{\alpha+\beta+\gamma}}{\Gamma(\alpha + \beta + \gamma + 1)} - (\eta + \mu)(\alpha + \beta + 1)\frac{t^{\alpha+2\beta+1}}{\Gamma(\alpha + 2\beta + 2)}$$
$$+ \delta(\beta + 1)(2\beta + 2)\frac{t^{3\beta+2}}{\Gamma(3\beta + 3)} + \eta(\beta + 2)(2\beta + 3)\frac{t^{3\beta+3}}{\Gamma(3\beta + 4)},$$

$$z_3(t) = -\delta\mu\frac{t^{\alpha+\beta+\gamma}}{\Gamma(\alpha + \beta + \gamma + 1)} - \mu\eta\frac{t^{\alpha+\beta+\gamma+1}}{\Gamma(\alpha + \beta + \gamma + 2)} + \frac{t^{\alpha+2\gamma+1}}{\Gamma(\alpha + 2\gamma + 2)}$$
$$\times \left\{ -\mu - (\eta + \mu)\frac{\Gamma(\alpha + \gamma + 2)}{\Gamma(\alpha + 1)\Gamma(\gamma + 2)} \right\} + \frac{t^{\alpha+2\gamma}}{\Gamma(\alpha + 2\gamma + 1)}$$
$$\times \left\{ -\mu^2(\delta - 1) - \mu(\eta + \mu)(\delta - 1)\frac{\Gamma(\alpha + \gamma + 1)}{\Gamma(\alpha + 1)\Gamma(\gamma + 1)} + \mu(\eta + \mu)(1 - \delta) \right\}$$
$$+ (\delta - 1)^2\frac{t^{3\gamma+1}}{\Gamma(3\gamma + 2)} + \mu(\delta - 1)^3\frac{t^{3\gamma}}{\Gamma(3\gamma + 1)}.$$

So up to third-order approximation, one may have the crisp solution

$$x(t) = x_0(t) + x_1(t) + x_2(t) + x_3(t),$$
$$y(t) = y_0(t) + y_1(t) + y_2(t) + y_3(t),$$

and

$$z(t) = z_0(t) + z_1(t) + z_2(t) + z_3(t).$$

TABLE 7.1 Uncertain (Fuzzy) and Crisp Solution Using HPM for $\alpha = \beta = \gamma = 1/3$

$t \rightarrow$		0	0.2	0.4	0.6	0.8	1
$r = 0$	$[\underline{x}(t), \overline{x}(t)]$	[0.4, 0.6]	[−0.4268, −0.5302]	[−0.6169, −0.8988]	[−0.8568, −1.3097]	[−1.2075, −1.8308]	[−1.7045, −2.5026]
	$[\underline{y}(t), \overline{y}(t)]$	[1.4, 1.6]	[1.1726, 1.2649]	[0.9350, 0.9229]	[0.7668, 0.6107]	[0.7378, 0.3901]	[0.9497, 0.3512]
	$[\underline{z}(t), \overline{z}(t)]$	[0, 0.2]	[0.0948, 0.0674]	[0.1355, −0.0495]	[0.0842, −0.2229]	[−0.0703, −0.4580]	[−0.3334, −0.7569]
$r = 0.5$	$[\underline{x}(t), \overline{x}(t)]$	[0.45, 0.55]	[−0.4559, −0.5077]	[−0.6950, −0.8363]	[−0.9824, −1.2093]	[−1.3804, −1.6927]	[−1.9261, −2.3259]
	$[\underline{y}(t), \overline{y}(t)]$	[1.45, 1.55]	[1.1972, 1.2433]	[0.9350, 0.9289]	[0.7323, 0.6542]	[0.6569, 0.4830]	[0.8076, 0.5083]
	$[\underline{z}(t), \overline{z}(t)]$	[0.05, 0.15]	[0.0793, 0.0646]	[0.0709, −0.0238]	[−0.0214, −0.1784]	[−0.2070, −0.4054]	[−0.4902, −0.7077]
$r = 1$	$[\underline{x}(t), \overline{x}(t)]$	[0.5, 0.5]	[−0.4829, −0.4829]	[−0.7682, −0.7682]	[−1.1000, −1.1000]	[−1.5424, −1.5424]	[−2.1335, −2.1335]
	$[\underline{y}(t), \overline{y}(t)]$	[1.5, 1.5]	[1.2207, 1.2207]	[0.9329, 0.9329]	[0.6948, 0.6948]	[0.5720, 0.5720]	[0.6604, 0.6604]
	$[\underline{z}(t), \overline{z}(t)]$	[0.1, 0.1]	[0.0689, 0.0689]	[0.0170, 0.0170]	[−0.1101, −0.1101]	[−0.3203, −0.3203]	[−0.6169, −0.6169]
Crisp solution	$x(t)$	0.5	−0.4829	−0.7682	−1.1000	−1.5424	−2.1335
	$y(t)$	1.5	1.2207	0.9329	0.6948	0.5720	0.6604
	$z(t)$	0.1	0.0689	0.0170	−0.1101	−0.3203	−0.6169

TABLE 7.2 Uncertain (Fuzzy) and Crisp Solution Using HPM for $\alpha = \beta = \gamma = 1/2$

$t \rightarrow$		0	0.2	0.4	0.6	0.8	1
$r = 0$	$[\underline{x}(t), \overline{x}(t)]$	[0.4, 0.6]	[−0.3399, −0.3168]	[−0.6383, −0.7329]	[−0.9013, −1.1297]	[−1.19801, −1.5787]	[−1.5748, −2.1278]
	$[\underline{y}(t), \overline{y}(t)]$	[1.4, 1.6]	[1.4267, 1.5583]	[1.3159, 1.3913]	[1.2010, 1.1858]	[1.1324, 0.9799]	[1.1866, 0.8361]
	$[\underline{z}(t), \overline{z}(t)]$	[0, 0.2]	[0.0595, 0.1463]	[0.1196, 0.1062]	[0.1372, 0.0077]	[0.0774, −0.1685]	[−0.0862, −0.4361]
$r = 0.5$	$[\underline{x}(t), \overline{x}(t)]$	[0.45, 0.55]	[−0.3346, −0.3231]	[−0.6648, −0.7122]	[−0.9645, −1.079]	[−1.3035, −1.4942]	[−1.7284, −2.0055]
	$[\underline{y}(t), \overline{y}(t)]$	[1.45, 1.55]	[1.4601, 1.5259]	[1.3362, 1.3739]	[1.1999, 1.1923]	[1.0983, 1.0221]	[1.1046, 0.9294]
	$[\underline{z}(t), \overline{z}(t)]$	[0.05, 0.15]	[0.0783, 0.1213]	[0.1082, 0.1005]	[0.0893, 0.0226]	[−0.0091, −0.1352]	[−0.2102, −0.3896]
$r = 1$	$[\underline{x}(t), \overline{x}(t)]$	[0.5, 0.5]	[−0.3291, −0.3291]	[−0.6894, −0.6894]	[−1.0238, −1.0238]	[−1.4024, −1.4024]	[−1.8721, −1.8721]
	$[\underline{y}(t), \overline{y}(t)]$	[1.5, 1.5]	[1.4931, 1.4931]	[1.35556, 1.35556]	[1.1969, 1.1969]	[1.0615, 1.0615]	[1.0189, 1.0189]
	$[\underline{z}(t), \overline{z}(t)]$	[0.1, 0.1]	[0.0987, 0.0987]	[0.1015, 0.1015]	[0.0504, 0.0504]	[−0.0811, −0.0811]	[−0.3129, −0.3129]
Crisp solution	$x(t)$	0.5	−0.3291	−0.6894	−1.0238	−1.4024	−1.8721
	$y(t)$	1.5	1.4931	1.3555	1.1969	1.0615	1.0189
	$z(t)$	0.1	0.0987	0.1015	0.0504	−0.0811	−0.3129

TABLE 7.3 Uncertain (Fuzzy) and Crisp Solution Using HPM for $\alpha = \beta = \gamma = 2/3$

$t \rightarrow$		0	0.2	0.4	0.6	0.8	1
$r = 0$	$[\underline{x}(t), \bar{x}(t)]$	[0.4, 0.6]	[−0.1745, −0.0979]	[−0.5367, −0.5376]	[−0.8623, −0.9526]	[−1.1917, −1.3922]	[−1.5580, −1.8942]
	$[\underline{y}(t), \bar{y}(t)]$	[1.4, 1.6]	[1.5194, 1.6649]	[1.5221, 1.6338]	[1.4979, 1.5586]	[1.4771, 1.4543]	[1.5079, 1.3525]
	$[\underline{z}(t), \bar{z}(t)]$	[0, 0.2]	[0.0478, 0.1711]	[0.1058, 0.1754]	[0.1533, 0.1466]	[0.1563, 0.0521]	[0.0800, −0.1352]
$r = 0.5$	$[\underline{x}(t), \bar{x}(t)]$	[0.45, 0.55]	[−0.1552, −0.1169]	[−0.5375, −0.5380]	[−0.8872, −0.9324]	[−1.2469, −1.3473]	[−1.651, −1.8194]
	$[\underline{y}(t), \bar{y}(t)]$	[1.45, 1.55]	[1.5559, 1.6287]	[1.5506, 1.6065]	[1.5144, 1.5448]	[1.4738, 1.4624]	[1.4728, 1.3951]
	$[\underline{z}(t), \bar{z}(t)]$	[0.05, 0.15]	[0.0771, 0.1387]	[0.1198, 0.1542]	[0.1443, 0.1399]	[0.1165, 0.0626]	[0.0035, −0.1070]
$r = 1$	$[\underline{x}(t), \bar{x}(t)]$	[0.5, 0.5]	[−0.1360, −0.1360]	[−0.5380, −0.5380]	[−0.9106, −0.9106]	[−1.2988, −1.2988]	[−1.7382, −1.7382]
	$[\underline{y}(t), \bar{y}(t)]$	[1.5, 1.5]	[1.5923, 1.5923]	[1.5788, 1.5788]	[1.5300, 1.5300]	[1.4689, 1.4689]	[1.4352, 1.4352]
	$[\underline{z}(t), \bar{z}(t)]$	[0.1, 0.1]	[0.1074, 0.1074]	[0.1358, 0.1358]	[0.1395, 0.1395]	[0.0846, 0.0846]	[−0.0598, −0.0598]
Crisp solution	$x(t)$	0.5	−0.1360	−0.5380	−0.9106	−1.2988	−1.7382
	$y(t)$	1.5	1.5923	1.57881	1.5300	1.4689	1.4352
	$z(t)$	0.1	0.1074	0.1358	0.1395	0.0846	−0.0598

7.4   NUMERICAL RESULTS

In this section, the numerical solution of fuzzy arbitrary-order Rossler's system using HPM has been presented. Few representative results with respect to various parameters involved in the corresponding equation are reported.

Fuzzy arbitrary (fractional)-order Rossler's system for $a(t) = t, b(t) = t, c(t) = 1$ with uncertain (fuzzy) initial condition as discussed earlier is solved by HPM. The solutions are given in Tables 7.1–7.3 for various values of $\alpha, \beta, \gamma, r,$ and t. By varying

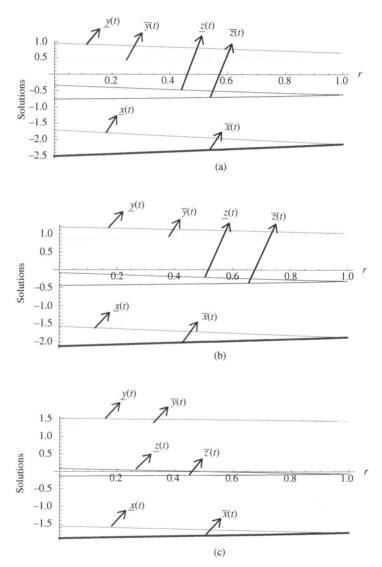

Figure 7.1   Uncertain (Fuzzy) solution of arbitrary order Rossler's system for $x(t), y(t)$ and $z(t)$ at $t = 1$ (a) $\alpha = \beta = \gamma = 1/3$ (b) $\alpha = \beta = \gamma = 1/2$ (c) $\alpha = \beta = \gamma = 2/3$

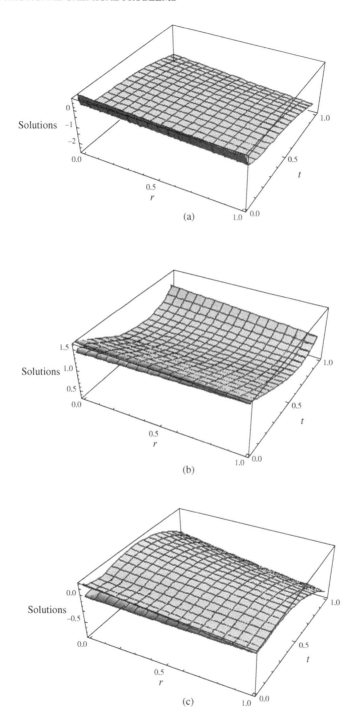

Figure 7.2 Uncertain (Fuzzy) solution of arbitrary-order Rossler's system for (a) $x(t)$, (b) $y(t)$, and (c) $z(t)$ where $t, r \in [0, 1]$ when $\alpha = \beta = \gamma = 1/3$

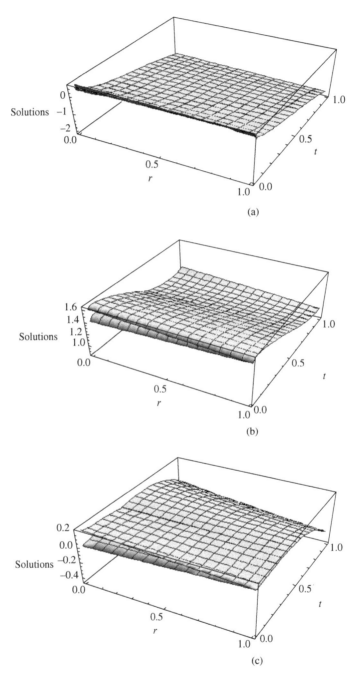

Figure 7.3 Uncertain (Fuzzy) solution of arbitrary-order Rossler's system for (a) $x(t)$, (b) $y(t)$, and (c) $z(t)$ where $t, r \in [0, 1]$ when $\alpha = \beta = \gamma = 1/2$

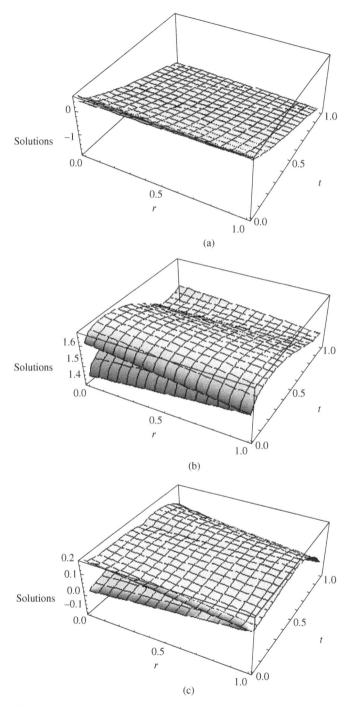

Figure 7.4 Uncertain (fuzzy) solution of arbitrary-order Rossler's system for (a) $x(t)$ (b) $y(t)$, and (c) $z(t)$ where $t, r \in [0, 1]$ when $\alpha = \beta = \gamma = 2/3$

t from 0 to 1 and keeping r constant for different types of fraction order, namely $\alpha = \beta = \gamma = 1/3, \alpha = \beta = \gamma = 1/2$, and $\alpha = \beta = \gamma = 2/3$, the numerical results are depicted in Tables 7.1–7.3, respectively. Also, these tables include results obtained by using HPM for crisp initial conditions.

Next, by taking $t = 1$ as constant and varying the value of r from 0 to 1, the obtained results are depicted in Figs. 7.1(a)–(c) for various arbitrary-order derivatives, namely $\alpha = \beta = \gamma = 1/3, \alpha = \beta = \gamma = 1/2$, and $\alpha = \beta = \gamma = 2/3$, respectively.

Now varying both t and r from 0 to 1 for $\alpha = \beta = \gamma = 1/3, \alpha = \beta = \gamma = 1/2$, and $\alpha = \beta = \gamma = 2/3$, the results are depicted, respectively, in Figs. 7.2–7.4. For the crisp initial condition, the solution of fractional order Rossler's system with crisp initial conditions $x(0) = \delta = 0.5, y(0) = \eta = 1.5$, and $y(0) = \mu = 0.1$ using HPM are depicted in Fig. 7.5.

As discussed earlier, for $r = 1$, the fuzzy initial condition converts into crisp initial value. As such, for crisp initial value, the obtained results by HPM exactly match with the HPM solution of fuzzy fractional-order Rossler's system for $r = 1$. Results are tabulated in Tables 7.1–7.3, and these are found to be in good agreement. The accuracy can be improved for all the cases by computing more number of terms in

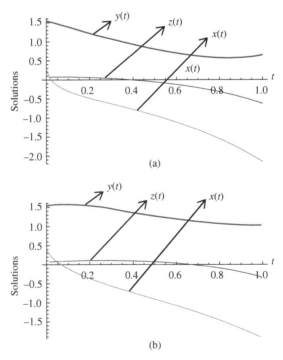

Figure 7.5   Crisp solution of arbitrary order Rossler's system for $x(t)$, $y(t)$ and $z(t)$ at $t = 1$ (a) $\alpha = \beta = \gamma = 1/3$ (b) $\alpha = \beta = \gamma = 1/2$ (c) $\alpha = \beta = \gamma = 2/3$ (d) $\alpha = \beta = \gamma = 1$

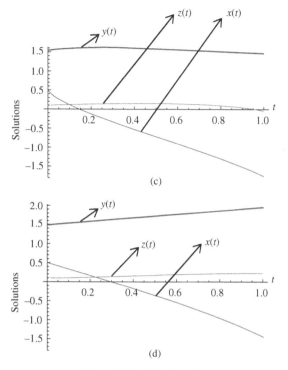

Figure 7.5 (*Continued*)

HPM for the solution. It is interesting to note that for all the cases, the lower and upper bounds of the fuzzy solutions are the same for $r = 1$ and also approximately equal to those of the crisp solution

BIBLIOGRAPHY

Agrawal RP, Lakshmikantham V, Nieto JJ. On the concept of solution for fractional differential equations with uncertainty. *Nonlinear Anal* 2010;**72**:2859–2862.

Arshad S, Lupulescu V. On the fractional differential equations with uncertainty. *Nonlinear Anal* 2011a;**74**:3685–3693.

Arshad S, Lupulescu V. Fractional differential equation with the fuzzy initial condition. *Electron J Differ Eqs* 2011b;**2011**:1–8.

Behera D, Chakraverty S. Numerical solution of fractionally damped beam by homotopy perturbation method. *Cent Eur J Phys* 2013;**11**:792–798.

Cano YC, Flores HR. Comparison between some approaches to solve fuzzy differential equations. *Fuzzy Sets Syst* 2009;**160**:1517–1527.

Chakraverty S, Behera D. Dynamic responses of fractionally damped mechanical system using homotopy perturbation method. *Alexandria Eng J* 2013;**53**:557–562.

Chen X, Liu B. Existence and uniqueness theorem for uncertain differential equations. *Fuzzy Optim Decis Making* 2010;**9**:69–81.

Deng H, Li T, Wang Q, Li H. A fractional-order hyperchaotic system and its synchronization. *Chaos Solitons Fractals* 2009;**41**:962–969.

Dubois D, Prade H. *Fuzzy Sets and Systems: Theory and Applications*. New York: Academic Press; 1980.

Hanss M. *Applied Fuzzy Arithmetic: An Introduction with Engineering Applications*. Berlin: Springer-Verlag; 2005.

He JH. Homotopy perturbation technique. *Comput Methods Appl Mech Eng* 1999;**178**: 257–262.

He JH. A coupling method of a homotopy technique and a perturbation technique for non-linear problems. *Int J Nonlin Mech* 2000;**35**:37–43.

He JH. Homotopy perturbation method for solving boundary value problems. *Phys Lett* 2006;**A350**:87–88.

Li C, Chen G. Chaos and hyperchaos in the fractional-order Rossler equations. *Phys A: Stat Mech Its Appl* 2004;**341**:55–61.

Miller KS, Ross B. *An Introduction to the Fractional Calculus and Fractional Differential Equations*. New York, NY: John Wiley & Sons, Inc.; 1993.

Moaddy K, Hashim I, Momani S. Non-standard finite difference schemes for solving fractional-order Rossler chaotic and hyperchaotic systems. *Comput Math Appl* 2011;**62**:1068–1074.

Oldham KB, Spanier J. *The Fractional Calculus*. New York, NY: Academic Press; 1974.

Petras I. *Fractional-Order Nonlinear Systems Modeling, Analysis and Simulation*. Heidelberg, Dordrecht, London, New York: Higher Education, Springer; 2011.

Podlubny I. *Fractional Differential Equations*. New York, NY: Academic Press; 1999.

Razminia A, Majd VJ, Baleanu D. Chaotic incommensurate fractional order Rossler system: active control and synchronization. *Adv Differ Eqs* 2011;**2011**:1–12.

Ross TJ. *Fuzzy Logic with Engineering Applications*. Wiley Student Edition; 2007.

Salahshour S, Allahviranloo T, Abbasbandy S. Solving Fuzzy Fractional Differential Equations by Fuzzy Laplace Transforms. *Commun Nonlinear Sci Numer Simulat* 2012;**17**: 1372–1381.

Samko SG, Kilbas AA, Marichev OI. *Fractional Integrals and Derivatives: Theory and Applications*. Langhorne, PA: Gordon and Breach Science Publishers; 1993.

Tapaswini S, Chakraverty S. Numerical solution of n-th order fuzzy linear differential equations by homotopy perturbation method. *Int J Comput Appl* 2013a;**64**:5–10.

Tapaswini S, Chakraverty S. Numerical solution of uncertain beam equations using double parametric form of fuzzy numbers. *Appl Comput Intell Soft Comput* 2013b;**2013**:1–8.

Wang Q. Homotopy perturbation method for fractional KdV equation. *Appl Math Comput* 2007;**190**:1795–1802.

Wang H, Liu Y. Existence Results for fractional fuzzy differential equations with finite delay. *Int Math Forum* 2011;**6**:2535–2538.

Wang Y, Yin X, Liu Y. Control Chaos in System with Fractional Order. *J Modern Phys* 2012;**3**:496–501.

Yuan L, Agrawal OP. A numerical scheme for dynamic systems containing fractional derivatives. *J Vib Acoust* 2002;**124**:321–324.

Zhang W, Zhou S, Li H, Zhu H. Chaos in a fractional-order Rossler system. *Chaos Solitons Fractals* 2009;**42**:1684–1691.

Zimmermann HJ. *Fuzzy Set Theory and Its Application*. Vol. **2001**. Boston/Dordrecht/London: Kluwer Academic Publishers; 2001.

8

FUZZY FRACTIONAL STRUCTURAL PROBLEMS

This chapter presents the solution of fuzzy fractionally damped structural systems. In this regard, both discrete and continuous systems are taken into consideration. First, in this chapter, the solution of imprecisely defined fractional-order discrete system, subject to unit step and impulse loads, has been discussed. A mechanical spring–mass system having fractional damping of order 1/2 with fuzzy initial condition is taken into consideration for this. Fuzziness appeared in the initial conditions are modeled through different types of convex, normalized fuzzy sets, namely triangular, trapezoidal, and Gaussian fuzzy numbers. Homotopy perturbation method (HPM) is used with fuzzy-based approach to obtain the uncertain response.

Next in this chapter, fuzzy fractionally damped continuous system, namely beam, has been studied using the double parametric form of fuzzy numbers subject to unit step and impulse loads. Triangular fuzzy numbers are used to represent the initial conditions. Using alpha cut, the corresponding beam equation is first converted to an interval-based equation. Next, it has been transformed to crisp form by applying double parametric form of fuzzy numbers. Here also HPM has been applied to obtain the fuzzy response.

- Problem 1

Fuzzy Arbitrary Order System: Fuzzy Fractional Differential Equations and Applications, First Edition.
Snehashish Chakraverty, Smita Tapaswini, and Diptiranjan Behera.
© 2016 John Wiley & Sons, Inc. Published 2016 by John Wiley & Sons, Inc.

8.1   FUZZY FRACTIONALLY DAMPED DISCRETE SYSTEM

A fuzzy fractionally damped single degree of freedom spring–mass system may be written as

$$mD^2 \widetilde{x}(t; r) + cD^\alpha \widetilde{x}(t; r) + k\widetilde{x}(t; r) = f(t) \tag{8.1}$$

where, m, c, and k represent the mass, damping, and stiffness coefficients, respectively. $f(t)$ is the externally applied force, and $D^\alpha \widetilde{x}(t; r)$ for $0 < \alpha < 1$, is the derivative of order α of the fuzzy displacement function $\widetilde{x}(t; r) = [\underline{x}(t; r), \overline{x}(t; r)]$. Here, $\widetilde{x}(t; r)$ is represented by r-cut form of fuzzy displacements. Although the coefficient α (known as the memory parameter) may take any value between 0 and 1, the value of 1/2 has been adopted here. This is because it has been shown that it describes the frequency dependence of the damping materials quite satisfactorily in the crisp fractional dynamic systems (Suarez and Shokooh, 1997; Yuan and Agrawal, 2002). Fuzzy initial displacements $\widetilde{x}(0)$ and initial velocity $v(0) = \widetilde{\dot{x}}(0)$ are taken as triangular, trapezoidal, and Gaussian fuzzy number, respectively, in Cases 1–3 as depicted in Table 8.1.

Through r-cut approach, the fuzzy initial conditions for Cases 1–3 given in Table 8.1 are then expressed as given in Table 8.2.

Equation (8.1) may now be written as

$$D^2 \widetilde{x}(t; r) + \frac{c}{m}D^{1/2}\widetilde{x}(t; r) + \frac{k}{m}\widetilde{x}(t; r) = \frac{f(t)}{m}. \tag{8.2}$$

According to HPM, we may construct a simple homotopy for an embedding parameter $p \in [0, 1]$ as follows:

$$(1 - p)(D^2 \widetilde{X}(t; r) - D^2 \widetilde{x}_0(t; r))$$

$$+ p\left(D^2 \widetilde{X}(t; r) + \frac{c}{m}D^{1/2}\widetilde{X}(t; r) + \frac{k}{m}\widetilde{X}(t; r) - \frac{f(t)}{m}\right) = 0, \tag{8.3}$$

or

$$D^2 \widetilde{X}(t; r) - D^2 \widetilde{x}_0(t; r) + p\left(D^2 \widetilde{x}_0(t; r) + \frac{c}{m}D^{1/2}\widetilde{X}(t; r) + \frac{k}{m}\widetilde{X}(t; r) - \frac{f(t)}{m}\right) = 0. \tag{8.4}$$

In the changing process from 0 to 1, for $p = 0$, Eq. (8.3) or (8.4) gives

$$(D^2 \widetilde{X}(t; r) - D^2 \widetilde{x}_0(t; r)) = 0,$$

and for $p = 1$, we have the original system

$$\left(D^2 \widetilde{X}(t; r) + \frac{c}{m}D^{1/2}\widetilde{X}(t; r) + \frac{k}{m}\widetilde{x}(t; r) - \frac{f(t)}{m}\right) = 0.$$

TABLE 8.1 Data for Fuzzy Initial Conditions

Initial Conditions	Case 1	Case 2	Case 3
$\widetilde{x}(0)$	(−0.1, 0, 0.1)	(−0.12, −0.06, 0.06, 0.12)	(0, 0.2, 0.2)
$v(0) = \dot{\widetilde{x}}(0)$	(−0.1, 0, 0.1)	(−0.12, −0.06, 0.06, 0.12)	(0, 0.2, 0.2)

TABLE 8.2 r-Cut Representations of Fuzzy Initial Conditions

Initial Conditions	Case 1	Case 2	Case 3
$\widetilde{x}(0; \alpha)$	[0.1r − 0.1, 0.1 − 0.1r]	[0.06r − 0.12, −0.06r + 0.12]	$[-0.2\sqrt{-2\log_e(r)}, 0.2\sqrt{-2\log_e(r)}]$
$v(0; \alpha)$ $= \dot{\widetilde{x}}(0; \alpha)$	[0.1r − 0.1, 0.1 − 0.1r]	[0.06r − 0.12, −0.06r + 0.12]	$[-0.2\sqrt{-2\log_e(r)}, 0.2\sqrt{-2\log_e(r)}]$

This is called deformation in topology.

$$(D^2 \widetilde{X}(t;r) - D^2 \widetilde{x}_0(t;r)) \text{ and } \left(D^2 \widetilde{X}(t;r) + \frac{c}{m} D^{1/2} \widetilde{X}(t;r) + \frac{k}{m} \widetilde{X}(t;r) - \frac{f(t)}{m} \right)$$

are called homotopic.

Next, we assume solution of Eq. (8.3) or (8.4) as a power series expansion in p as

$$\widetilde{X}(t;r) = \widetilde{X}_0(t;r) + p\widetilde{X}_1(t;r) + p^2 \widetilde{X}_2(t;r) + p^3 \widetilde{X}_3(t;r) + \cdots, \tag{8.5}$$

where $\widetilde{X}_i(t;r)$, $i = 0, 1, 2, \ldots$ are functions yet to be determined. As per HPM, substituting Eq. (8.5) into Eq. (8.3) or (8.4), and equating the terms with the identical power of p, we can obtain a series of equations of the form

$$p^1 : \ D^2 \widetilde{X}_1(t;r) + D^2 \widetilde{x}_0(t;r) + \frac{c}{m} D^{1/2} \widetilde{X}_0(t;r) + \frac{k}{m} \widetilde{X}_0(t;r) - \frac{f(t)}{m} = 0,$$

$$p^2 : \ D^2 \widetilde{X}_2(t;r) + \frac{c}{m} D^{1/2} \widetilde{X}_1(t;r) + \frac{k}{m} \widetilde{X}_1(t;r) = 0,$$

$$p^3 : \ D^2 \widetilde{X}_3(t;r) + \frac{c}{m} D^{1/2} \widetilde{X}_2(t;r) + \frac{k}{m} \widetilde{X}_2(t;r) = 0,$$

$$p^4 : \ D^2 \widetilde{X}_4(t;r) + \frac{c}{m} D^{1/2} \widetilde{X}_3(t;r) + \frac{k}{m} \widetilde{X}_3(t;r) = 0,$$

$$p^5 : \ D^2 \widetilde{X}_5(t;r) + \frac{c}{m} D^{1/2} \widetilde{X}_4(t;r) + \frac{k}{m} \widetilde{X}_4(t;r) = 0,$$

$$p^6 : \ D^2 \widetilde{X}_6(t;r) + \frac{c}{m} D^{1/2} \widetilde{X}_5(t;r) + \frac{k}{m} \widetilde{X}_5(t;r) = 0, \tag{8.6}$$

and so on.

Applying the operator L_{tt}^{-1} (the inverse operator of $D^2 = d^2/dt^2$) on both sides of Eq. (8.6), one may get the approximate solution $\tilde{x}(t; r) = \lim\limits_{p \to 1} \tilde{X}(t; r)$, which can be expressed as

$$\tilde{x}(t; r) = \tilde{X}_0(t; r) + \tilde{X}_1(t; r) + \tilde{X}_2(t; r) + \tilde{X}_3(t; r) + \cdots. \tag{8.7}$$

We can write the aforementioned expression equivalently as

$$[\underline{x}(t; r), \overline{x}(t; r)] = \sum_{n=0}^{\infty} \tilde{X}_n(t; r), \quad \text{where } \tilde{X}_n(t; r) = [\underline{X}_n(t; r), \overline{X}_n(t; r)].$$

Hence, the lower and upper bounds of the solution in parametric form are given as $\underline{x}(t; r) = \sum_{n=0}^{\infty} \underline{X}_n(t; r)$ and $\overline{x}(t; r) = \sum_{n=0}^{\infty} \overline{X}_n(t; r)$, respectively.

8.2 UNCERTAIN RESPONSE ANALYSIS

In this section, the beam has been analyzed with respect to unit step and impulse loading as follows for the different cases as given in Tables 8.1 and 8.2.

8.2.1 Uncertain Step Function Response

The unit step load has been considered as $f(t) = u(t)$, where $u(t)$ is the Heaviside function.

- Solution for Case 1
  For triangular fuzzy initial conditions, we have

$$\underline{X}_0(t; r) = (0.1r - 0.1),$$

$$\overline{X}_0(t; r) = (0.1 - 0.1r),$$

$$\underline{X}_1(t; r) = -\frac{t^2}{2}\left(\frac{k}{m}(0.1r - 0.1) - \frac{u(t)}{m}\right),$$

$$\overline{X}_1(t; r) = -\frac{t^2}{2}\left(\frac{k}{m}(0.1 - 0.1r) - \frac{u(t)}{m}\right),$$

$$\underline{X}_2(t; r) = (0.1r - 0.1)\frac{k}{m}\left(\frac{c}{m}\frac{t^{7/2}}{\Gamma(9/2)} + \frac{k}{m}\frac{t^4}{\Gamma(5)}\right)$$

$$+ u(t)\left(-\frac{c}{m^2}\frac{t^{7/2}}{\Gamma(9/2)} + \frac{k}{m^2}\frac{t^4}{\Gamma(5)}\right),$$

$$\overline{X}_2(t; r) = (0.1 - 0.1r)\frac{k}{m}\left(\frac{c}{m}\frac{t^{7/2}}{\Gamma(9/2)} + \frac{k}{m}\frac{t^4}{\Gamma(5)}\right)$$

$$+ u(t) \left(-\frac{c}{m^2} \frac{t^{7/2}}{\Gamma(9/2)} + \frac{k}{m^2} \frac{t^4}{\Gamma(5)} \right),$$

$$\underline{X}_3(t; r) = (0.1r - 0.1) \frac{k}{m} \left(-\frac{c^2}{m^2} \frac{t^5}{\Gamma(6)} - \frac{2kc}{m^2} \frac{t^{11/2}}{\Gamma(13/2)} - \frac{k^2}{m^2} \frac{t^6}{\Gamma(7)} \right)$$

$$+ u(t) \left(\frac{c^2}{m^3} \frac{t^5}{\Gamma(6)} + \frac{2kc}{m^3} \frac{t^{11/2}}{\Gamma(13/2)} + \frac{k^2}{m^3} \frac{t^6}{\Gamma(7)} \right),$$

$$\overline{X}_3(t; r) = (0.1 - 0.1r) \frac{k}{m} \left(-\frac{c^2}{m^2} \frac{t^5}{\Gamma(6)} - \frac{2kc}{m^2} \frac{t^{11/2}}{\Gamma(13/2)} - \frac{k^2}{m^2} \frac{t^6}{\Gamma(7)} \right)$$

$$+ u(t) \left(\frac{c^2}{m^3} \frac{t^5}{\Gamma(6)} + \frac{2kc}{m^3} \frac{t^{11/2}}{\Gamma(13/2)} + \frac{k^2}{m^3} \frac{t^6}{\Gamma(7)} \right),$$

and so on.

Substituting these in Eq. (8.7), we may get the approximate solution of $\widetilde{x}(t)$. Accordingly, the bounds of the general solution may be

$$\underline{x}(t; r) = (0.1r - 0.1) + \left(\sum_{n=0}^{\infty} \frac{(-1)^n}{n!} \left(\frac{k}{m} \right)^n t^{2(n+1)} \sum_{j=0}^{\infty} \left(\frac{-c}{m} \right)^j \frac{(j+n)! t^{3j/2}}{j! \Gamma\left(\frac{3j}{2} + 2n + 3 \right)} \right)$$

$$\times \left(\frac{u(t)}{m} - \frac{k}{m} (0.1r - 0.1) \right)$$

$$= (0.1r - 0.1) + \left(\sum_{n=0}^{\infty} \frac{(-1)^n}{n!} \left(\frac{k}{m} \right)^n t^{2(n+1)} E_{3/2, n/2+3}^n \left(\frac{-c}{m} t^{3/2} \right) \right)$$

$$\times \left(\frac{u(t)}{m} - \frac{k}{m} (0.1r - 0.1) \right) \tag{8.8}$$

and

$$\overline{x}(t; r) = (0.1 - 0.1r) + \left(\sum_{n=0}^{\infty} \frac{(-1)^n}{n!} \left(\frac{k}{m} \right)^n t^{2(n+1)} \sum_{j=0}^{\infty} \left(\frac{-c}{m} \right)^j \frac{(j+n)! t^{3j/2}}{j! \Gamma\left(\frac{3j}{2} + 2n + 3 \right)} \right)$$

$$\times \left(\frac{u(t)}{m} - \frac{k}{m} (0.1 - 0.1r) \right)$$

$$= (0.1 - 0.1r) + \left(\sum_{n=0}^{\infty} \frac{(-1)^n}{n!} \left(\frac{k}{m} \right)^n t^{2(n+1)} E_{3/2, n/2+3}^n \left(\frac{-c}{m} t^{3/2} \right) \right)$$

$$\times \left(\frac{u(t)}{m} - \frac{k}{m} (0.1 - 0.1r) \right). \tag{8.9}$$

- Solution for Case 2

  For trapezoidal fuzzy initial condition, the general solution can be represented as

$$\underline{x}(t;r) = (0.06r - 0.12) + \left(\sum_{n=0}^{\infty} \frac{(-1)^n}{n!} \left(\frac{k}{m}\right)^n t^{2(n+1)} \sum_{j=0}^{\infty} \left(\frac{-c}{m}\right)^j \frac{(j+n)!\, t^{3j/2}}{j!\,\Gamma\left(\frac{3j}{2} + 2n + 3\right)} \right)$$

$$\times \left(\frac{u(t)}{m} - \frac{k}{m}(0.06r - 0.12) \right)$$

$$= (0.06r - 0.12) + \left(\sum_{n=0}^{\infty} \frac{(-1)^n}{n!} \left(\frac{k}{m}\right)^n t^{2(n+1)} E^n_{3/2, n/2+3} \left(\frac{-c}{m} t^{3/2}\right) \right)$$

$$\times \left(\frac{u(t)}{m} - \frac{k}{m}(0.06r - 0.12) \right), \tag{8.10}$$

and

$$\overline{x}(t;r) = (-0.06r + 0.12)$$

$$+ \left(\sum_{n=0}^{\infty} \frac{(-1)^n}{n!} \left(\frac{k}{m}\right)^n t^{2(n+1)} \sum_{j=0}^{\infty} \left(\frac{-c}{m}\right)^j \frac{(j+n)!\, t^{3j/2}}{j!\,\Gamma\left(\frac{3j}{2} + 2n + 3\right)} \right)$$

$$\times \left(\frac{u(t)}{m} - \frac{k}{m}(0.06r + 0.12) \right)$$

$$= (0.06r + 0.12) + \left(\sum_{n=0}^{\infty} \frac{(-1)^n}{n!} \left(\frac{k}{m}\right)^n t^{2(n+1)} E^n_{3/2, n/2+3} \left(\frac{-c}{m} t^{3/2}\right) \right)$$

$$\times \left(\frac{u(t)}{m} - \frac{k}{m}(-0.06r + 0.12) \right). \tag{8.11}$$

- Solution for Case 3

  Similarly for Gaussian fuzzy initial condition, one may have the general solution as

$$\underline{x}(t;r) = (-0.2\sqrt{-2\log_e(r)})$$

$$+ \left(\sum_{n=0}^{\infty} \frac{(-1)^n}{n!} \left(\frac{k}{m}\right)^n t^{2(n+1)} \sum_{j=0}^{\infty} \left(\frac{-c}{m}\right)^j \frac{(j+n)!\, t^{3j/2}}{j!\,\Gamma\left(\frac{3j}{2} + 2n + 3\right)} \right)$$

$$\times \left(\frac{u(t)}{m} - \frac{k}{m}(-0.2\sqrt{-2\log_e(r)}) \right)$$

$$= (-0.2\sqrt{-2\log_e(r)}) + \left(\sum_{n=0}^{\infty} \frac{(-1)^n}{n!} \left(\frac{k}{m}\right)^n t^{2(n+1)} E^n_{3/2,r/2+3} \left(\frac{-c}{m} t^{3/2}\right) \right)$$

$$\times \left(\frac{u(t)}{m} - \frac{k}{m}(-0.2\sqrt{-2\log_e(r)}) \right). \tag{8.12}$$

and

$$\bar{x}(t;r) = (0.2\sqrt{-2\log_e(r)})$$

$$+ \left[\sum_{n=0}^{\infty} \frac{(-1)^n}{n!} \left(\frac{k}{m}\right)^n t^{2(n+1)} \sum_{j=0}^{\infty} \left(\frac{-c}{m}\right)^j \frac{(j+n)! t^{3j/2}}{j! \Gamma\left(\frac{3j}{2} + 2n + 3\right)} \right]$$

$$\times \left(\frac{u(t)}{m} - \frac{k}{m}(0.2\sqrt{-2\log_e(r)}) \right)$$

$$= (0.2\sqrt{-2\log_e(r)}) + \left(\sum_{n=0}^{\infty} \frac{(-1)^n}{n!} \left(\frac{k}{m}\right)^n t^{2(n+1)} E^n_{3/2,n/2+3} \left(\frac{-c}{m} t^{3/2}\right) \right)$$

$$\times \left(\frac{u(t)}{m} - \frac{k}{m}(0.2\sqrt{-2\log_e(r)}) \right). \tag{8.13}$$

8.2.2 Uncertain Impulse Function Response

We have considered the response subject to unit impulsive load, namely $f(t) = \delta(t)$, where $\delta(t)$ is the unit impulse function.

- Solution for Case 1

  For triangular fuzzy initial conditions, we have

$$\underline{X}_0(t;r) = (0.1r - 0.1),$$

$$\overline{X}_0(t;r) = (0.1 - 0.1r),$$

$$\underline{X}_1(t;r) = -(0.1r - 0.1)\frac{k}{m}\frac{t^2}{2} + \frac{t}{m},$$

$$\overline{X}_1(t;r) = -(0.1 - 0.1r)\frac{k}{m}\frac{t^2}{2} + \frac{t}{m},$$

$$\underline{X}_2(t;r) = (0.1r - 0.1)\frac{k}{m}\left(\frac{c}{m}\frac{t^{7/2}}{\Gamma(9/2)} + \frac{k}{m}\frac{t^4}{\Gamma(5)}\right) - \frac{c}{m^2}\frac{t^{5/2}}{\Gamma(7/2)} - \frac{k}{m^2}\frac{t^3}{\Gamma(4)},$$

$$\overline{X}_2(t;r) = (0.1 - 0.1r)\frac{k}{m}\left(\frac{c}{m}\frac{t^{7/2}}{\Gamma(9/2)} + \frac{k}{m}\frac{t^4}{\Gamma(5)}\right) - \frac{c}{m^2}\frac{t^{5/2}}{\Gamma(7/2)} - \frac{k}{m^2}\frac{t^3}{\Gamma(4)},$$

$$\underline{X}_3(t;r) = -(0.1r - 0.1)\frac{k}{m}\left(\frac{c^2}{m^2}\frac{t^5}{\Gamma(6)} + \frac{2kc}{m^2}\frac{t^{9/2}}{\Gamma(11/2)} + \frac{k^2}{m^2}\frac{t^6}{\Gamma(7)}\right)\frac{c^2}{m^3}\frac{t^4}{\Gamma(5)}$$

$$+ \frac{2kc}{m^3} \frac{t^{9/2}}{\Gamma(11/2)} + \frac{k^2}{m^3} \frac{t^5}{\Gamma(6)},$$

$$\overline{X}_3(t\,;r) = -(0.1 - 0.1r)\frac{k}{m}\left(\frac{c^2}{m^2}\frac{t^5}{\Gamma(6)} + \frac{2kc}{m^2}\frac{t^{9/2}}{\Gamma(11/2)} + \frac{k^2}{m^2}\frac{t^6}{\Gamma(7)}\right)\frac{c^2}{m^3}\frac{t^4}{\Gamma(5)}$$

$$+ \frac{2kc}{m^3}\frac{t^{9/2}}{\Gamma(11/2)} + \frac{k^2}{m^3}\frac{t^5}{\Gamma(6)},$$

and so on.

Substituting these in Eq. (8.7), we may get the approximate solution of $\widetilde{x}(t)$. Accordingly, the general solution may be written as

$$\underline{x}(t\,;r) = (0.1r - 0.1)$$

$$\times\left(1 - \frac{k}{m}\left[\sum_{n=0}^{\infty}\frac{(-1)^n}{n!}\left(\frac{k}{m}\right)^n t^{2(n+1)}\sum_{j=0}^{\infty}\left(\frac{-c}{m}\right)^j\frac{(j+n)!\,t^{3j/2}}{j!\Gamma\left(\frac{3j}{2}+2n+3\right)}\right]\right)$$

$$+\frac{1}{m}\sum_{n=0}^{\infty}\frac{(-1)^n}{n!}\left(\frac{k}{m}\right)^n t^{2n+1}\sum_{j=0}^{\infty}\left(\frac{-c}{m}\right)^j\frac{(j+n)!\,t^{3j/2}}{j!\Gamma\left(\frac{3j}{2}+2n+2\right)}$$

$$= (0.1r - 0.1)\left(1 - \frac{k}{m}\left(\sum_{n=0}^{\infty}\frac{(-1)^n}{n!}\left(\frac{k}{m}\right)^n t^{2(n+1)}E_{3/2,n/2+3}^n\left(\frac{-c}{m}t^{3/2}\right)\right)\right)$$

$$+\frac{1}{m}\sum_{n=0}^{\infty}\frac{(-1)^n}{n!}\left(\frac{k}{m}\right)^n t^{2n+1}E_{3/2,n/2+2}^n\left(\frac{-c}{m}t^{3/2}\right) \qquad (8.14)$$

and

$$\overline{x}(t\,;r) = (0.1 - 0.1r)$$

$$\times\left(1 - \frac{k}{m}\left[\sum_{n=0}^{\infty}\frac{(-1)^n}{n!}\left(\frac{k}{m}\right)^n t^{2(n+1)}\sum_{j=0}^{\infty}\left(\frac{-c}{m}\right)^j\frac{(j+n)!\,t^{3j/2}}{j!\Gamma\left(\frac{3j}{2}+2n+3\right)}\right]\right)$$

$$+\frac{1}{m}\sum_{n=0}^{\infty}\frac{(-1)^r}{r!}\left(\frac{k}{m}\right)^n t^{2n+1}\sum_{j=0}^{\infty}\left(\frac{-c}{m}\right)^j\frac{(j+n)!\,t^{3j/2}}{j!\Gamma\left(\frac{3j}{2}+2n+2\right)}$$

$$= (0.1 - 0.1r)\left(1 - \frac{k}{m}\left(\sum_{n=0}^{\infty}\frac{(-1)^n}{n!}\left(\frac{k}{m}\right)^n t^{2(n+1)}E_{3/2,r/2+3}^n\left(\frac{-c}{m}t^{3/2}\right)\right)\right)$$

$$+\frac{1}{m}\sum_{n=0}^{\infty}\frac{(-1)^n}{n!}\left(\frac{k}{m}\right)^n t^{2n+1}E_{3/2,n/2+2}^n\left(\frac{-c}{m}t^{3/2}\right) \qquad (8.15)$$

- Solution for Case 2

For trapezoidal fuzzy initial condition, the general solution can be represented as

$$\underline{x}(t\,;r) = (0.06r - 0.12)$$

$$\times \left(1 - \frac{k}{m} \left[\sum_{n=0}^{\infty} \frac{(-1)^n}{n!} \left(\frac{k}{m}\right)^n t^{2(n+1)} \sum_{j=0}^{\infty} \left(\frac{-c}{m}\right)^j \frac{(j+n)!\,t^{3j/2}}{j!\,\Gamma\left(\frac{3j}{2}+2n+3\right)} \right] \right)$$

$$+ \frac{1}{m} \sum_{n=0}^{\infty} \frac{(-1)^n}{n!} \left(\frac{k}{m}\right)^n t^{2n+1} \sum_{j=0}^{\infty} \left(\frac{-c}{m}\right)^j \frac{(j+n)!\,t^{3j/2}}{j!\,\Gamma\left(\frac{3j}{2}+2n+2\right)}$$

$$= (0.06r - 0.12)$$

$$\times \left(1 - \frac{k}{m} \left(\sum_{n=0}^{\infty} \frac{(-1)^n}{n!} \left(\frac{k}{m}\right)^n t^{2(n+1)} E^n_{3/2,n/2+3} \left(\frac{-c}{m}t^{3/2}\right) \right) \right)$$

$$+ \frac{1}{m} \sum_{n=0}^{\infty} \frac{(-1)^n}{n!} \left(\frac{k}{m}\right)^n t^{2n+1} E^n_{3/2,n/2+2} \left(\frac{-c}{m}t^{3/2}\right) \qquad (8.16)$$

and

$$\overline{x}(t\,;r) = (-0.06r + 0.12)$$

$$\times \left(1 - \frac{k}{m} \left[\sum_{n=0}^{\infty} \frac{(-1)^n}{n!} \left(\frac{k}{m}\right)^n t^{2(n+1)} \sum_{j=0}^{\infty} \left(\frac{-c}{m}\right)^j \frac{(j+n)!\,t^{3j/2}}{j!\,\Gamma\left(\frac{3j}{2}+2n+3\right)} \right] \right)$$

$$+ \frac{1}{m} \sum_{n=0}^{\infty} \frac{(-1)^r}{r!} \left(\frac{k}{m}\right)^n t^{2n+1} \sum_{j=0}^{\infty} \left(\frac{-c}{m}\right)^j \frac{(j+n)!\,t^{3j/2}}{j!\,\Gamma\left(\frac{3j}{2}+2n+2\right)}$$

$$= (-0.06r + 0.12)$$

$$\times \left(1 - \frac{k}{m} \left(\sum_{n=0}^{\infty} \frac{(-1)^n}{n!} \left(\frac{k}{m}\right)^n t^{2(n+1)} E^n_{3/2,r/2+3} \left(\frac{-c}{m}t^{3/2}\right) \right) \right)$$

$$+ \frac{1}{m} \sum_{n=0}^{\infty} \frac{(-1)^n}{n!} \left(\frac{k}{m}\right)^n t^{2n+1} E^n_{3/2,n/2+2} \left(\frac{-c}{m}t^{3/2}\right) \qquad (8.17)$$

- Solution for Case 3

  Similarly for Gaussian fuzzy initial condition, one may have the general solution as

$$\underline{x}(t\,;r) = -0.2\sqrt{-2\log_e(r)}$$

$$\times \left(1 - \frac{k}{m} \left[\sum_{n=0}^{\infty} \frac{(-1)^n}{n!} \left(\frac{k}{m}\right)^n t^{2(n+1)} \sum_{j=0}^{\infty} \left(\frac{-c}{m}\right)^j \frac{(j+n)!\, t^{3j/2}}{j!\,\Gamma\left(\frac{3j}{2}+2n+3\right)} \right] \right)$$

$$+ \frac{1}{m} \sum_{n=0}^{\infty} \frac{(-1)^n}{n!} \left(\frac{k}{m}\right)^n t^{2n+1} \sum_{j=0}^{\infty} \left(\frac{-c}{m}\right)^j \frac{(j+n)!\, t^{3j/2}}{j!\,\Gamma\left(\frac{3j}{2}+2n+2\right)}$$

$$= -0.2\sqrt{-2\log_e(r)}$$

$$\times \left(1 - \frac{k}{m} \left(\sum_{n=0}^{\infty} \frac{(-1)^n}{n!} \left(\frac{k}{m}\right)^n t^{2(n+1)} E_{3/2,n/2+3}^{n} \left(\frac{-c}{m} t^{3/2}\right) \right) \right)$$

$$+ \frac{1}{m} \sum_{n=0}^{\infty} \frac{(-1)^n}{n!} \left(\frac{k}{m}\right)^n t^{2n+1} E_{3/2,n/2+2}^{n} \left(\frac{-c}{m} t^{3/2}\right), \qquad (8.18)$$

and

$$\overline{x}(t\,;r) = 0.2\sqrt{-2\log_e(r)}$$

$$\times \left(1 - \frac{k}{m} \left[\sum_{n=0}^{\infty} \frac{(-1)^n}{n!} \left(\frac{k}{m}\right)^n t^{2(n+1)} \sum_{j=0}^{\infty} \left(\frac{-c}{m}\right)^j \frac{(j+n)!\, t^{3j/2}}{j!\,\Gamma\left(\frac{3j}{2}+2n+3\right)} \right] \right)$$

$$+ \frac{1}{m} \sum_{n=0}^{\infty} \frac{(-1)^n}{n!} \left(\frac{k}{m}\right)^n t^{2n+1} \sum_{j=0}^{\infty} \left(\frac{-c}{m}\right)^j \frac{(j+n)!\, t^{3j/2}}{j!\,\Gamma\left(\frac{3j}{2}+2n+2\right)}$$

$$= 0.2\sqrt{-2\log_e(r)}$$

$$\times \left(1 - \frac{k}{m} \left(\sum_{n=0}^{\infty} \frac{(-1)^n}{n!} \left(\frac{k}{m}\right)^n t^{2(n+1)} E_{3/2,n/2+3}^{n} \left(\frac{-c}{m} t^{3/2}\right) \right) \right)$$

$$+ \frac{1}{m} \sum_{n=0}^{\infty} \frac{(-1)^n}{n!} \left(\frac{k}{m}\right)^n t^{2n+1} E_{3/2,n/2+2}^{n} \left(\frac{-c}{m} t^{3/2}\right). \qquad (8.19)$$

8.3 NUMERICAL RESULTS

For numerical simulations, we use the notations of the parameters $\omega_n = \sqrt{k/m}$, $\eta = c/2m\omega_n^{3/2}$ and value of $m = 1$, where ω_n is the natural frequency and η is the damping ratio.

8.3.1   Case Studies for Uncertain Step Function Response

Equations (8.8)–(8.13) provide the desired expressions for the considered loading condition. Figure 8.1 shows the triangular fuzzy response for Case 1 with natural frequencies $\omega_n = 5$ rad/s (Fig. 8.1(a)), $\omega_n = 10$ rad/s (Fig. 8.1(b)) and damping ratio $\eta = 0.05$ for Eqs. (8.8) and (8.9). Similar simulations have been done for Cases 2 and 3. Accordingly, the trapezoidal and Gaussian fuzzy responses are depicted in Figs. 8.2 and 8.3.

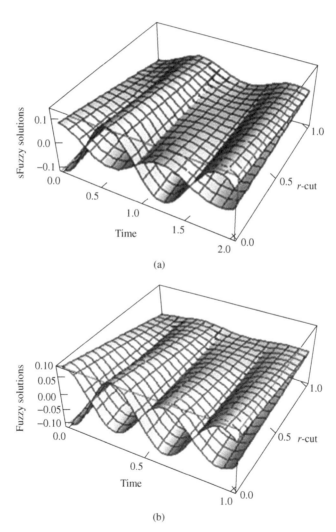

(a)

(b)

Figure 8.1   Triangular fuzzy response subject to unit step load for Case 1 with natural frequency (a) $\omega_n = 5$ rad/s, (b) $\omega_n = 10$ rad/s, and damping ratio $\eta = 0.05$

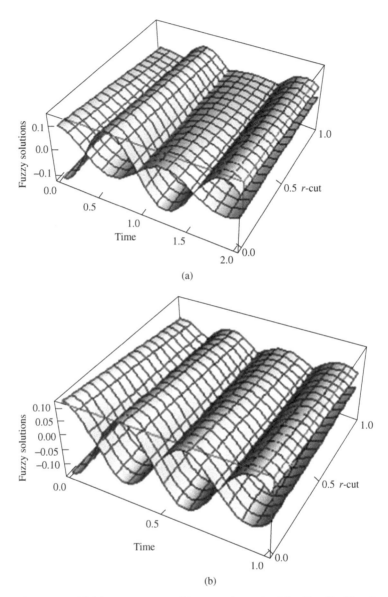

Figure 8.2 Trapezoidal fuzzy response subject to unit step load for Case 2 with natural frequency (a) $\omega_n = 5$ rad/s, (b) $\omega_n = 10$ rad/s, and damping ratio $\eta = 0.05$

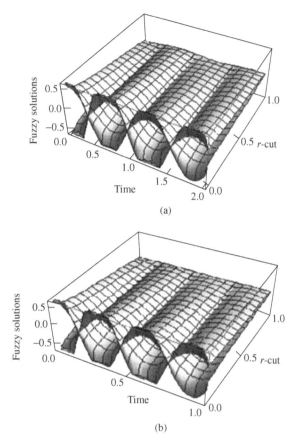

(a)

(b)

Figure 8.3 Gaussian fuzzy response subject to unit step load for Case 3 with natural frequency (a) $\omega_n = 5$ rad/s, (b) $\omega_n = 10$ rad/s, and damping ratio $\eta = 0.05$

In Case 1, the triplet number (a, b, c) defines a triangular membership function, where a and c are the lower and upper bounds at $\alpha = 0$ and b is the nominal (crisp) value at $r = 1$. Hence, for $r = 0$ (Fig. 8.4(a)) and $r = 1$ (Fig. 8.4(b)) along with the crisp solution by Podlubny (1999) for crisp initial condition, these are depicted in Fig. 8.4. Similarly, Figs. 8.5 and 8.6 represent the uncertain-but-bounded (interval) solutions for Cases 2 and 3. Here in these cases, $\omega_n = 5$ rad/s and $\eta = 0.05$ are considered. Similar interval responses are shown in Figs. 8.7–8.9 with $\omega_n = 10$ rad/s, and $\eta = 0.05$.

One may see from Figs. 8.1 and 8.3 for Cases 1 and 3 that lower and upper bounds of the fuzzy displacements coincide for $r = 1$, as the fuzzy initial conditions convert to the crisp one (Chakraverty and Behera, 2013). Also, it is interesting to note from Figs. 8.4–8.9 that for $r = 0$, interval bounds contain the crisp solution. Also, interval solution bounds coincide with the crisp solutions for Cases 1 and 3.

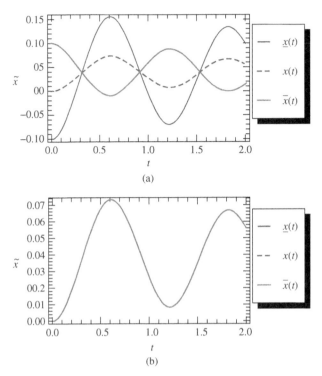

Figure 8.4 Uncertain but bounded (interval) response subject to unit step load for Case 1 when (a) $r = 0$, (b) $r = 1$ with crisp analytical solution (- - -) by Podlubny (1999) where natural frequency $\omega_n = 5$ rad/s and damping ratio $\eta = 0.05$

8.3.2 Case Studies for Uncertain Impulse Function Response

Depending on the values of natural frequency ω_n and damping ratio η, different cases have been studied. First, the numerical values of the natural frequency $\omega_n = 5$ rad/s and damping ratio $\eta = 0.05$ are taken. Next, natural frequency $\omega_n = 10$ rad/s with damping ratio $\eta = 0.05$ with unit impulse load is considered for the oscillation. With these parametric values obtained, fuzzy displacements are depicted in Figs. 8.10–8.12. Also, one can see from Figs. 8.10 and 8.12 for Cases 1 and 3 that lower and upper bounds of the fuzzy displacements coincide for $r = 1$. This is because the fuzzy initial conditions again convert to a crisp one (Chakraverty and Behera, 2013).

HPM with fuzzy-based approach has successfully been applied to obtain the uncertain solution of an imprecisely defined fractionally damped spring–mass mechanical system subject to a unit step and impulse load, where the fractional derivative is considered of order 1/2.

- Problem 2

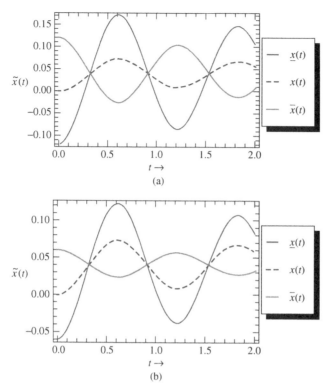

Figure 8.5 Uncertain but bounded (interval) response subject to unit step load for Case 2 when (a) $r = 0$, (b) $r = 1$ with crisp analytical solution (- - -) by Podlubny (1999) where natural frequency $\omega_n = 5$ rad/s and damping ratio $\eta = 0.05$

8.4 FUZZY FRACTIONALLY DAMPED CONTINUOUS SYSTEM

Let us consider a fuzzy linear differential equation, which describes the dynamics of the aforementioned system, that is, viscoelastic beam with damping as an arbitrary fractional derivative of order α

$$\rho A \frac{\partial^2 \tilde{v}}{\partial t^2} + c \frac{\partial^\alpha \tilde{v}}{\partial t^\alpha} + EI \frac{\partial^4 \tilde{v}}{\partial x^4} = F(x, t). \tag{8.20}$$

Equation (8.20) may be written as

$$\frac{\partial^2 \tilde{v}}{\partial t^2} + \frac{c}{\rho A} \frac{\partial^\alpha \tilde{v}}{\partial t^\alpha} + \frac{EI}{\rho A} \frac{\partial^4 \tilde{v}}{\partial x^4} = \frac{F(x, t)}{\rho A}, \tag{8.21}$$

where ρ, A, c, E, and I are the mass density, cross-sectional area, damping coefficients per unit length, Young's modulus of elasticity, and moment of inertia of the beam. $F(x, t)$ is the externally applied force and $\tilde{v}(x, t)$ is the transverse fuzzy

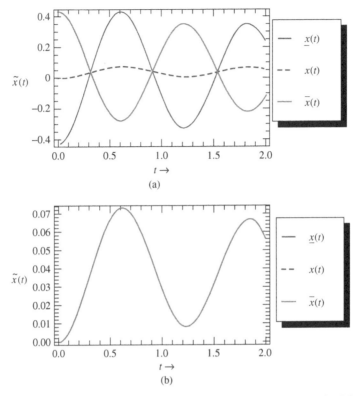

Figure 8.6 Uncertain but bounded (interval) response subject to unit step load for Case 3 when (a) $r = 0$, (b) $r = 1$ with crisp analytical solution (- - -) by Podlubny (1999) where natural frequency $\omega_n = 5$ rad/s and damping ratio $\eta = 0.05$

displacement. $\partial^\alpha / \partial t^\alpha$ is the fractional derivative of order $\alpha \in (0, \ 1)$ of the fuzzy displacement function $\widetilde{v}(x, \ t)$. Initial conditions are considered as fuzzy, namely $\widetilde{v}(0) = \widetilde{v}'(0) = (-0.1, \ 0, \ 0.1)$.

As per the single parametric form, we may write Eq. (8.21) as

$$\left[\frac{\partial^2 \underline{v}(x, t; r)}{\partial t^2}, \frac{\partial^2 \overline{v}(x, t; r)}{\partial t^2}\right] + \frac{c}{\rho A}\left[\frac{\partial^\alpha \underline{v}(x, t; r)}{\partial t^\alpha}, \frac{\partial^\alpha \overline{v}(x, t; r)}{\partial t^\alpha}\right]$$

$$+ \frac{EI}{\rho A}\left[\frac{\partial^4 \underline{v}(x, t; r)}{\partial x^4}, \frac{\partial^4 \overline{v}(x, t; r)}{\partial x^4}\right] = \frac{F(x, t)}{\rho A} \tag{8.22}$$

subject to fuzzy initial conditions

$$[\underline{v}(x, 0; r), \overline{v}(x, 0; r)]$$
$$= [\underline{v}'(x, 0; r), \overline{v}'(x, 0; r)][0.1r - 0.1, 0.1 - 0.1r], \quad \text{where } r \in [0, 1].$$

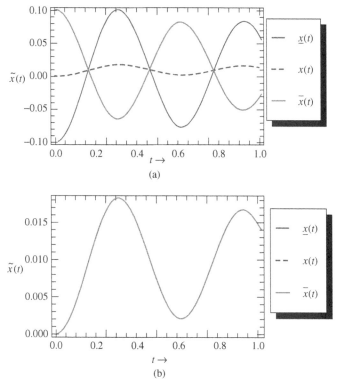

Figure 8.7 Uncertain but bounded (interval) response subject to unit step load for Case 1 when (a) $r = 0$, (b) $r = 1$ with crisp analytical solution (- - -) by Podlubny (1999) where natural frequency $\omega_n = 10$ rad/s and damping ratio $\eta = 0.05$

Next, using the double parametric form (as discussed in Chapter 1), Eq. (8.22) can be expressed as

$$
\left\{ \beta \left(\frac{\partial^2 \overline{v}(x,t;r)}{\partial t^2} - \frac{\partial^2 \underline{v}(x,t;r)}{\partial t^2} \right) + \frac{\partial^2 \underline{v}(x,t;r)}{\partial t^2} \right\}
$$
$$
+ \frac{c}{\rho A} \left\{ \beta \left(\frac{\partial^\alpha \overline{v}(x,t;r)}{\partial t^\alpha} - \frac{\partial^\alpha \underline{v}(x,t;r)}{\partial t^\alpha} \right) + \frac{\partial^\alpha \underline{v}(x,t;r)}{\partial t^\alpha} \right\}
$$
$$
+ \frac{EI}{\rho A} \left\{ \beta \left(\frac{\partial^4 \overline{v}(x,t;r)}{\partial x^4} - \frac{\partial^4 \underline{v}(x,t;r)}{\partial x^4} \right) + \frac{\partial^4 \underline{v}(x,t;r)}{\partial x^4} \right\} = \frac{F(x,t)}{\rho A} \quad (8.23)
$$

subject to the initial conditions

$$
\{ \beta(\overline{v}(x,0;r) - \underline{v}(x,0;r)) + \underline{v}(x,0;r) \} == \{ \beta(0.2 - 0.2r) + (0.1r - 0.1) \},
$$

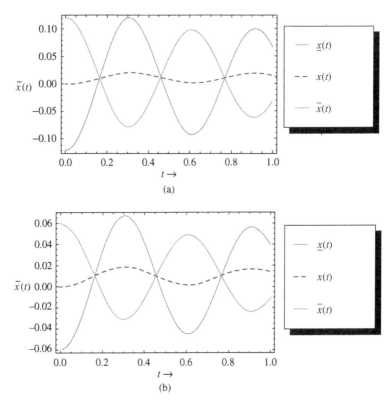

Figure 8.8 Uncertain but bounded (interval) response subject to unit step load for Case 2 when (a) $r = 0$, (b) $r = 1$ with crisp analytical solution (- - -) by Podlubny (1999) where natural frequency $\omega_n = 10$ rad/s and damping ratio $\eta = 0.05$

and

$$\{\beta(\overline{v}'(x, 0; r) - \underline{v}'(x, 0; r)) + \underline{v}'(x, 0; r)\} = \{\beta(0.2 - 0.2r) + (0.1r - 0.1)\},$$

where $\beta \in [0, 1]$.

Let us now denote

$$\left\{\beta\left(\frac{\partial^2 \overline{v}(x, t; r)}{\partial t^2} - \frac{\partial^2 \underline{v}(x, t; r)}{\partial t^2}\right) + \frac{\partial^2 \underline{v}(x, t; r)}{\partial t^2}\right\} = \frac{\partial^2 \widetilde{v}(x, t; r, \beta)}{\partial t^2},$$

$$\left\{\beta\left(\frac{\partial^\alpha \overline{v}(x, t; r)}{\partial t^\alpha} - \frac{\partial^\alpha \underline{v}(x, t; r)}{\partial t^\alpha}\right) + \frac{\partial^\alpha \underline{v}(x, t; r)}{\partial t^\alpha}\right\} = \frac{\partial^\alpha \widetilde{v}(x, t; r, \beta)}{\partial t^\alpha},$$

$$\left\{\beta\left(\frac{\partial^4 \overline{v}(x, t; r)}{\partial x^4} - \frac{\partial^4 \underline{v}(x, t; r)}{\partial x^4}\right) + \frac{\partial^4 \underline{v}(x, t; r)}{\partial x^4}\right\} = \frac{\partial^4 \widetilde{v}(x, t; r, \beta)}{\partial x^4},$$

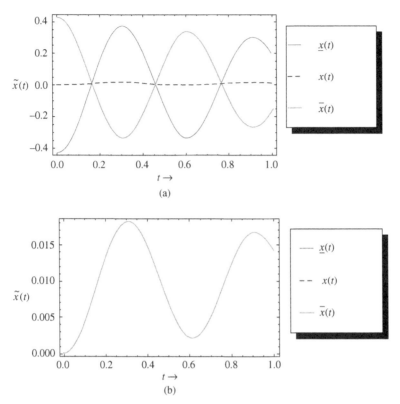

Figure 8.9 Uncertain but bounded (interval) response subject to unit step load for Case 3 when (a) $r = 0$, (b) $r = 1$ with crisp analytical solution (- - -) by Podlubny (1999) where natural frequency $\omega_n = 10$ rad/s and damping ratio $\eta = 0.05$

$$\{\beta(\widetilde{v}(x, 0; r) - \underline{v}(0; r)) + \underline{v}(x, 0; r)\} = \widetilde{v}(x, 0; r, \beta) \quad \text{and}$$

$$\{\beta(\widetilde{v}'(x, 0; r) - \underline{v}'(x, 0; r)) + \underline{v}'(x, 0; r)\} = \widetilde{v}'(x, 0; r, \beta).$$

Substituting these values in Eq. (8.23), we get

$$\frac{\partial^2 \widetilde{v}(x, t; r, \beta)}{\partial t^2} + \frac{c}{\rho A} \frac{\partial^\alpha \widetilde{v}(x, t; r, \beta)}{\partial t^\alpha} + \frac{EI}{\rho A} \frac{\partial^4 \widetilde{v}(x, t; r, \beta)}{\partial x^4} = \frac{F(x, t)}{\rho A}, \tag{8.24}$$

with initial conditions $\widetilde{v}(x, 0; r, \beta) = \widetilde{v}'(x, 0; r, \beta) = \{\beta(0.2 - 0.2r) + (0.1r - 0.1)\}$.

Now, Eq. (8.24) has been solved using HPM. According to HPM, we may construct a simple homotopy for an embedding parameter $p \in [0, 1]$, as follows:

$$(1 - p)\frac{\partial^2 \widetilde{v}(x, t; r, \beta)}{\partial t^2}$$

$$+ p\left[\frac{\partial^2 \widetilde{v}(x, t; r, \beta)}{\partial t^2} + \frac{c}{\rho A} \frac{\partial^\alpha \widetilde{v}(x, t; r, \beta)}{\partial t^\alpha} + \frac{EI}{\rho A} \frac{\partial^4 \widetilde{v}(x, t; r, \beta)}{\partial x^4} - \frac{F(x, t)}{\rho A}\right] = 0, \tag{8.25}$$

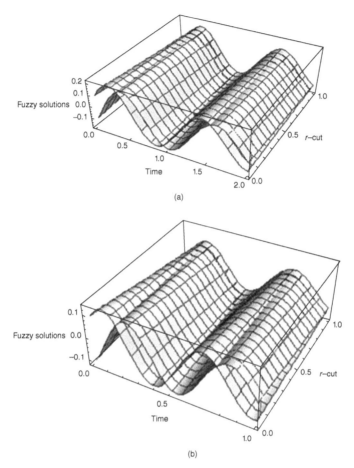

(a)

(b)

Figure 8.10   Triangular fuzzy response subject to unit impulse load for Case 1 with natural
frequency (a) $\omega_n = 5$ rad/s, (b) $\omega_n = 10$ rad/s and damping ratio $\eta = 0.05$

or

$$\frac{\partial^2 \widetilde{v}(x,t;r,\beta)}{\partial t^2} + p\left[\frac{c}{\rho A}\frac{\partial^\alpha \widetilde{v}(x,t;r,\beta)}{\partial t^\alpha} + \frac{EI}{\rho A}\frac{\partial^4 \widetilde{v}(x,t;r,\beta)}{\partial x^4} - \frac{F(x,t)}{\rho A}\right] = 0. \quad (8.26)$$

For $p = 0$, Eqs. (8.25) and (8.26) become a linear equation, that is,

$$\frac{\partial^2 \widetilde{v}(x,t;\alpha,\beta)}{\partial t^2} = 0,$$

which is easy to solve. For $p = 1$, Eqs. (8.25) and (8.26) turn out to be the same as
the original equation (8.24). This is called deformation in topology.

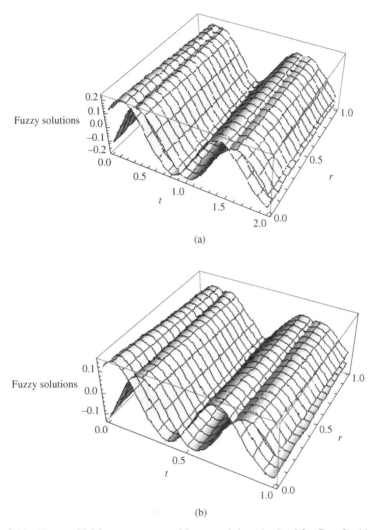

Figure 8.11 Trapezoidal fuzzy response subject to unit impulse load for Case 2 with natural frequency (a) $\omega_n = 5$ rad/s, (b) $\omega_n = 10$ rad/s and damping ratio $\eta = 0.05$

We assume the solution of Eq. (8.25) or (8.26) as a power series expansion in p as

$$\widetilde{v}(x,t;r,\beta) = \widetilde{v}_0(x,t;r,\beta) + p\widetilde{v}_1(x,t;r,\beta) + p^2\widetilde{v}_2(x,t;r,\beta) + p^3\widetilde{v}_3(x,t;r,\beta) + \cdots,$$
$$(8.27)$$

where, $\widetilde{v}_i(x,t;r,\beta)$ for $i = 0, 1, 2, 3, \ldots$ are functions yet to be determined.

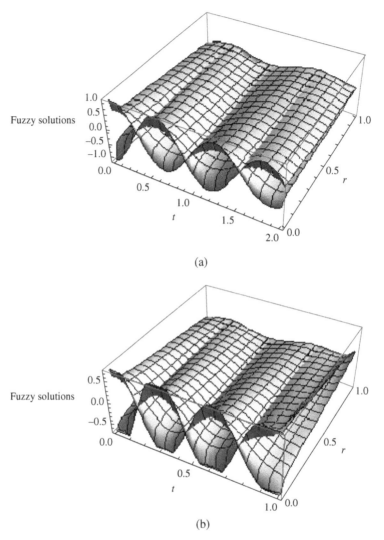

(a)

(b)

Figure 8.12 Gaussian fuzzy response subject to unit impulse load for Case 3 with natural frequency (a) $\omega_n = 5$ rad/s, (b) $\omega_n = 10$ rad/s and damping ratio $\eta = 0.05$

Substituting Eq. (8.27) into Eq. (8.25) or (8.26) and equating the terms with the identical powers of p, we have

$$p^0 : \ \frac{\partial^2 \widetilde{v}_0(x, t; r, \beta)}{\partial t^2} = 0, \tag{8.28}$$

$$p^1 : \frac{\partial^2 \tilde{v}_1(x,t;r,\beta)}{\partial t^2} + \frac{c}{\rho A} \frac{\partial^\alpha \tilde{v}_0(x,t;r,\beta)}{\partial t^\alpha} + \frac{EI}{\rho A} \frac{\partial^4 \tilde{v}_0(x,t;r,\beta)}{\partial x^4} - \frac{F(x,t)}{\rho A} = 0,$$

(8.29)

$$p^2 : \frac{\partial^2 \tilde{v}_2(x,t;r,\beta)}{\partial t^2} + \frac{c}{\rho A} \frac{\partial^\alpha \tilde{v}_1(x,t;r,\beta)}{\partial t^\alpha} + \frac{EI}{\rho A} \frac{\partial^4 \tilde{v}_1(x,t;r,\beta)}{\partial x^4} = 0,$$

(8.30)

$$p^3 : \frac{\partial^2 \tilde{v}_3(x,t;r,\beta)}{\partial t^2} + \frac{c}{\rho A} \frac{\partial^\lambda \tilde{v}_2(x,t;r,\beta)}{\partial t^\lambda} + \frac{EI}{\rho A} \frac{\partial^4 \tilde{v}_2(x,t;r,\beta)}{\partial x^4} = 0,$$

(8.31)

$$p^4 : \frac{\partial^2 \tilde{v}_4(x,t;r,\beta)}{\partial t^2} + \frac{c}{\rho A} \frac{\partial^\alpha \tilde{v}_3(x,t;r,\beta)}{\partial t^\alpha} + \frac{EI}{\rho A} \frac{\partial^4 \tilde{v}_3(x,t;r,\beta)}{\partial x^4} = 0,$$

(8.32)

and so on.

Choosing initial approximation

$$v(x,0;\alpha,\beta) = v'(x,0;\alpha,\beta) = \{\beta(0.2 - 0.2r) + (0.1r - 0.1)\}$$

and applying the inverse operator L_{tt}^{-1} (which is the inverse of the operator $L_{tt} = \partial^2/\partial t^2$) on both sides of each Eqs. (8.28)–(8.32), one may obtain the following equations:

$$\tilde{v}_0(x,t;r,\beta) = \tilde{v}'(x,0;r,\beta)t + \tilde{v}(x,0;r,\beta),$$

(8.33)

$$\tilde{v}_1(x,t;r,\beta) = L_{tt}^{-1} \left(-\frac{c}{\rho A} \frac{\partial^\alpha \tilde{v}_0(x,t;r,\beta)}{\partial t^\alpha} - \frac{EI}{\rho A} \frac{\partial^4 \tilde{v}_0(x,t;r,\beta)}{\partial x^4} + \frac{F(x,t)}{\rho A} \right),$$

(8.34)

$$\tilde{v}_2(x,t;r,\beta) = L_{tt}^{-1} \left(-\frac{c}{\rho A} \frac{\partial^\alpha \tilde{v}_1(x,t;r,\beta)}{\partial t^\alpha} - \frac{EI}{\rho A} \frac{\partial^4 \tilde{v}_1(x,t;r,\beta)}{\partial x^4} \right),$$

(8.35)

$$\tilde{v}_3(x,t;r,\beta) = L_{tt}^{-1} \left(-\frac{c}{\rho A} \frac{\partial^\alpha v_2(x,t;r,\beta)}{\partial t^\alpha} - \frac{EI}{\rho A} \frac{\partial^4 \tilde{v}_2(x,t;r,\beta)}{\partial x^4} \right),$$

(8.36)

$$\tilde{v}_4(x,t;r,\beta) = L_{tt}^{-1} \left(-\frac{c}{\rho A} \frac{\partial^\alpha \tilde{v}_3(x,t;r,\beta)}{\partial t^\alpha} - \frac{EI}{\rho A} \frac{\partial^4 \tilde{v}_3(x,t;r,\beta)}{\partial x^4} \right),$$

(8.37)

and so on.

Substituting these terms in Eq. (8.27) with, $p \to 1$, one may get the approximate solution of Eq. (8.24) as follows.

$$\tilde{v}(x,t;r,\beta) = \tilde{v}_0(x,t;r,\beta) + p\tilde{v}_1(x,t;r,\beta) + p^2 \tilde{v}_2(x,t;r,\beta) + p^3 \tilde{v}_3(x,t;r,\beta) + \cdots,$$

To obtain the lower and upper bounds of the solution in single parametric form, we may substitute $\beta = 0$ and 1, respectively. These may be represented as $v(x,t;r,0) = \underline{v}(x,t;r)$ and $v(x,t;r,1) = \bar{v}(x,t;r)$.

8.5   UNCERTAIN RESPONSE ANALYSIS

Let us consider the external applied force $F(x, t)$ as

$$F(x, t) = f(x)g(t),$$

where $f(x)$ is a specified space-dependent deterministic function, and $g(t)$ is a time-dependent process. In the following section, we examine the fuzzy response of the dynamic system (8.24) subject to unit step and impulse loading conditions.

8.5.1   Unit step Function Response

We now consider the response of the fuzzy fractionally damped beam subject to a unit step load of the form $g(t) = Bu(t)$ where $u(t)$ is the Heaviside function and B is a constant. By using HPM, we have

$$\tilde{v}_0(x, t; r, \beta) = \{\beta(0.2 - 0.2r) + (0.1r - 0.1)\}(1 + t), \tag{8.38}$$

$$\tilde{v}_1(x, t; r, \beta) = \{\beta(0.2 - 0.2r) + (0.1r - 0.1)\}\left(-\frac{c}{\rho A}\frac{t^{3-\alpha}}{\Gamma(4 - \alpha)}\right) + \frac{fBt^2}{2\rho A}, \tag{8.39}$$

$$\tilde{v}_2(x, t; r, \beta) = \{\beta(0.2 - 0.2r) + (0.1r - 0.1)\}\left(\frac{c^2}{\rho^2 A^2}\frac{t^{5-2\alpha}}{\Gamma(6 - 2\alpha)}\right) - \frac{fBc}{\rho^2 A^2}\frac{t^{4-\alpha}}{\Gamma(5 - \alpha)}$$

$$- \frac{EIBf^4}{\rho^2 A^2}\frac{t^4}{\Gamma(5)}, \tag{8.40}$$

$$\tilde{v}_3(x, t; r, \beta) = \{\beta(0.2 - 0.2r) + (0.1r - 0.1)\}\left(-\frac{c^3}{\rho^3 A^3}\frac{t^{7-3\alpha}}{\Gamma(8 - 3\alpha)}\right)$$

$$+ \frac{fBc^2}{\rho^3 A^3}\frac{t^{6-2\alpha}}{\Gamma(7 - 2\alpha)} + \frac{2EIBcf^4}{\rho^3 A^3}\frac{t^{6-\alpha}}{\Gamma(7 - \alpha)} + \frac{E^2 I^2 Bf^8}{\rho^3 A^3}\frac{t^6}{\Gamma(7)}, \tag{8.41}$$

$$\tilde{v}_4(x, t; r, \beta) = \{\beta(0.2 - 0.2r) + (0.1r - 0.1)\}\left(-\frac{c^4}{\rho^4 A^4}\frac{t^{9-4\alpha}}{\Gamma(10 - 4\alpha)}\right)$$

$$- \frac{fBc^3}{\rho^4 A^4}\frac{t^{8-3\alpha}}{\Gamma(9 - 3\alpha)} - \frac{3EIBc^2 f^4}{\rho^4 A^4}\frac{t^{8-2\alpha}}{\Gamma(9 - 2\alpha)}$$

$$- \frac{3E^2 I^2 Bcf^8}{\rho^4 A^4}\frac{t^{8-\alpha}}{\Gamma(9 - \alpha)} - \frac{3E^3 I^3 Bf^{12}}{\rho^4 A^4}\frac{t^8}{\Gamma(9)}, \tag{8.42}$$

and so on, where $f^{(i)} = \partial^i f / \partial x^i$. The solution can be written in general form as

$$\tilde{v}(x, t; r, \beta) = \{\beta(0.2 - 0.2r) + (0.1r - 0.1)\}\left\{1 + \sum_{k=0}^{\infty}\frac{t^{(2-\alpha)k+1}}{\Gamma((2 - \alpha)k + 2)}\right\}$$

$$+ \frac{B}{\rho A}\sum_{n=0}^{\infty}\frac{(-1)^n}{n!}\left(\frac{EI}{\rho A}\right)^n f^{(4n)}t^{2(n+1)}\sum_{j=0}^{\infty}\left(\frac{-c}{\rho A}\right)^j\frac{(j + n)!t^{(2-\alpha)j}}{j!\Gamma((2 - \alpha)j + 2n + 3)}. \tag{8.43}$$

As discussed earlier, the solution bound in single parametric form may be obtained by putting $\beta = 0$ and 1. This may be represented as

$$\widetilde{v}(x,t;r,0) = \underline{v}(x,t,r) \text{ and } \widetilde{v}(x,t,r,1) = \overline{v}(x,t,r),$$

where

$$\underline{v}(x,t;r,0) = \frac{B}{\rho A} \sum_{n=0}^{\infty} \frac{(-1)^n}{n!} \left(\frac{EI}{\rho A}\right)^n f^{(4n)} t^{2(n+1)} \sum_{j=0}^{\infty} \left(\frac{-c}{\rho A}\right)^j \frac{(j+n)! t^{(2-\alpha)j}}{j! \Gamma((2-\alpha)j + 2n + 3)}$$

(8.44)

and

$$\overline{v}(x,t;r,1) = \frac{B}{\rho A} \sum_{n=0}^{\infty} \frac{(-1)^n}{n!} \left(\frac{EI}{\rho A}\right)^n f^{(4n)} t^{2(n+1)} \sum_{j=0}^{\infty} \left(\frac{-c}{\rho A}\right)^j \frac{(j+n)! t^{(2-\alpha)j}}{j! \Gamma((2-\alpha)j + 2n + 3)}.$$

(8.45)

8.5.2 Unit Impulse Function Response

In this section, we study the response of the beam subject to unit impulse load of the form $g(t) = \delta(t)$ where $\delta(t)$ is the unit impulse function. Using HPM in this case again, we have

$$\widetilde{v}_0(x,t;r,\beta) = \{\beta(0.2 - 0.2r) + (0.1r - 0.1)\}(1+t),$$ (8.46)

$$\widetilde{v}_1(x,t;r,\beta) = \{\beta(0.2 - 0.2r) + (0.1r - 0.1)\} \left(-\frac{c}{\rho A} \frac{t^{3-\alpha}}{\Gamma(4-\alpha)}\right) + \frac{fBt}{\rho A},$$ (8.47)

$$\widetilde{v}_2(x,t;r,\beta) = \{\beta(0.2 - 0.2r) + (0.1r - 0.1)\} \left(\frac{c^2}{\rho^2 A^2} \frac{t^{5-2\alpha}}{\Gamma(6-2\alpha)}\right)$$

$$- \frac{fBc}{\rho^2 A^2} \frac{t^{3-\alpha}}{\Gamma(4-\alpha)} - \frac{EIBf^4}{\rho^2 A^2} \frac{t^3}{\Gamma(4)},$$ (8.48)

$$\widetilde{v}_3(x,t;r,\beta) = \{\beta(0.2 - 0.2r) + (0.1r - 0.1)\} \left(-\frac{c^3}{\rho^3 A^3} \frac{t^{7-3\alpha}}{\Gamma(8-3\alpha)}\right)$$

$$+ \frac{fBc^2}{\rho^3 A^3} \frac{t^{5-2\alpha}}{\Gamma(6-2\alpha)} + \frac{2EIBcf^4}{\rho^3 A^3} \frac{t^{5-\alpha}}{\Gamma(6-\alpha)} + \frac{E^2 I^2 Bf^8}{\rho^3 A^3} \frac{t^5}{\Gamma(6)},$$ (8.49)

and so on, where $f^{(i)} = \partial^i f / \partial x^i$. Hence, the solution can be written in general form as

$$\widetilde{v}(x,t;r,\beta) = \{\beta(0.2 - 0.2r) + (0.1r - 0.1)\} \left\{1 + \sum_{k=0}^{\infty} \frac{t^{(2-\alpha)k+1}}{\Gamma((2-\alpha)k + 2)}\right\}$$

$$+ \frac{1}{\rho A} \sum_{n=0}^{\infty} \frac{(-1)^n}{n!} \left(\frac{EI}{\rho A}\right)^n f^{(4n)} t^{2n+1} \sum_{j=0}^{\infty} \left(\frac{-c}{\rho A}\right)^j \frac{(j+n)! t^{(2-\alpha)j}}{j! \Gamma((2-\alpha)j + 2n + 2)}.$$ (8.50)

Again to obtain the solution bounds in single parametric form, we may put $\beta = 0$ and 1 to get the lower and upper bounds of the solution, respectively, as

$$\widetilde{v}(x, t; r, 0) = \underline{v}(x, t, r) \quad \text{and} \quad v(x, t, \alpha, 1) = \bar{v}(x, t, \alpha) \quad \widetilde{v}(x, t, r, 1) = \bar{v}(x, t, r)$$

where

$$\underline{v}(x, t; r, 0) = \frac{B}{\rho A} \sum_{n=0}^{\infty} \frac{(-1)^n}{n!} \left(\frac{EI}{\rho A} \right)^n f^{(4n)} t^{2n+1} \sum_{j=0}^{\infty} \left(\frac{-c}{\rho A} \right)^j \frac{(j+n)! t^{(2-\alpha)j}}{j! \Gamma((2-\alpha)j + 2n + 2)},$$
(8.51)

and

$$\bar{v}(x, t; r, 1) = \frac{B}{\rho A} \sum_{n=0}^{\infty} \frac{(-1)^n}{n!} \left(\frac{EI}{\rho A} \right)^n f^{(4n)} t^{2n+1} \sum_{j=0}^{\infty} \left(\frac{-c}{\rho A} \right)^j \frac{(j+n)! t^{(2-\alpha)j}}{j! \Gamma((2-\alpha)j + 2n + 2)}.$$
(8.52)

8.6 NUMERICAL RESULTS

Numerical results corresponding to the discussed loads have been considered in this section. Equations (8.43) and (8.50) provide the desired expressions for the considered loading condition. We have assumed a simply supported beam, hence one may have

$$f(x) = \sin \left(\frac{\pi x}{L} \right).$$

For numerical simulations, let us denote c/m and $EI/\rho A$, respectively, as $2\eta\omega^{3/2}$ and ω^2 where ω is the natural frequency and η is the damping ratio. The values of the parameters are taken as $B = 1$, $\rho A = 1$, $L = \pi$, $x = 1/2$, and $m = 1$.

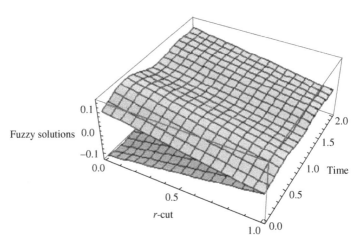

Figure 8.13   Fuzzy unit step response for $\omega = 5$ rad/s, $\eta = 0.5$, and $\alpha = 0.2$

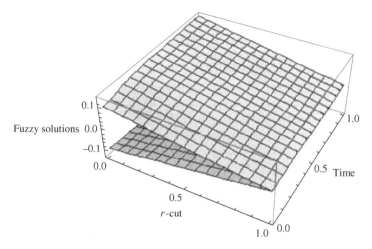

Figure 8.14 Fuzzy unit step response for $\omega = 10$ rad/s, $\eta = 0.05$, and $\alpha = 0.5$

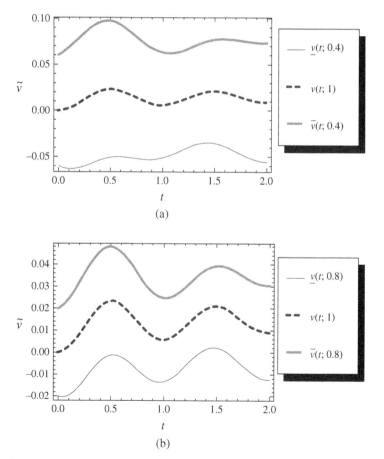

Figure 8.15 Interval unit step response for (a) $r = 0.4$ (b) $r = 0.8$ with $\omega = 5$ rad/s, $\eta = 0.5$ $\alpha = 0.2$, and $r = 1$

8.6.1   Case Studies for Fuzzy Unit Step Response

Depending on the natural frequency ω, damping ratio η, and arbitrary-order fractional derivative α subject to unit step load, two different cases have been considered as follows. In the first case, the numerical values of the parameters are taken as $\omega = 5\,\text{rad/s}$, $\eta = 0.5$, and $\alpha = 0.2$. In the second case, $\omega = 10$ rad/s, $\eta = 0.05$, and $\alpha = 0.5$ have been considered. For first and second cases, the obtained fuzzy responses with respect to time are depicted in Figs. 8.13 and 8.14, respectively.

Figures 8.15 and 8.16 give the effects of interval unit step responses for the particular membership r. For $r = 1$, the lower and upper bounds of the solution coincide with each other and are denoted as $\underline{v}(t; 1) = \overline{v}(t; 1) = v(t; 1)$, which is actually the crisp solution given in Behera and Chakraverty (2013). Figure 8.15 represents the interval solution for $r = 0.4$ and 0.8 with $r = 1$ for the first case. Similarly, Fig. 8.16 cites the results for the second case with $r = 1$.

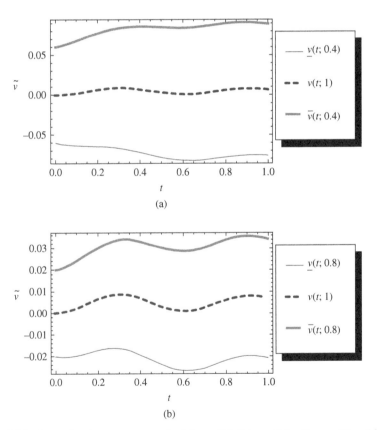

Figure 8.16   Interval unit step response for (a) $r = 0.4$, (b) $r = 0.8$ with $\omega = 10$ rad/s, $\eta = 0.05$, $\alpha = 0.5$, and $r = 1$

We now vary the fractional-order derivative with the same parametric values as considered for Fig. 8.15. As such, Figs. 8.17 and 8.18 present the interval unit step responses for $\alpha = 0.5$ and 0.8, respectively.

From Figs. 8.16–8.18, it can be seen that the uncertain width of the solution gradually decreases by increasing the value α. One may also observe from Figs. 8.15, 8.17, and 8.18 that the oscillation of the uncertain bounds of the unit step response gradually decreases by increasing the order of the fractional derivative.

8.6.2   Case Studies for Fuzzy Unit Impulse Response

Depending on the system parameters, namely natural frequency ω, damping ratio η, and arbitrary-order fractional derivative α, four different cases have been considered as follows subjected to unit impulse load.

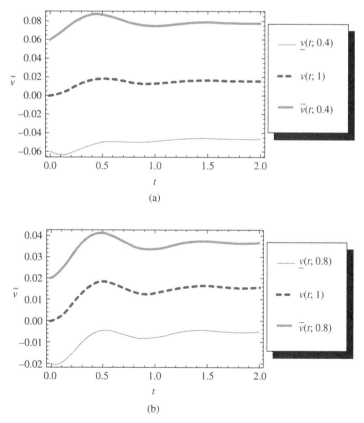

Figure 8.17   Interval unit step response for (a) $r = 0.4$, (b) $r = 0.8$ with $\omega = 5$ rad/s, $\eta = 0.5$, $\alpha = 0.5$, and $r = 1$

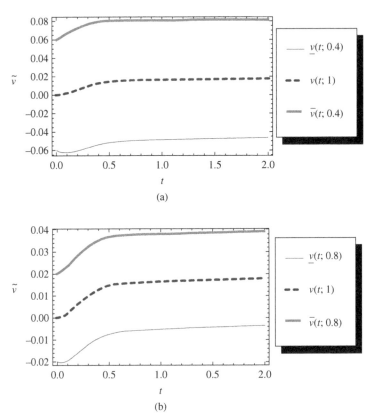

(a)

(b)

Figure 8.18 Interval unit step response for (a) $r = 0.4$, (b) $r = 0.8$ with $\omega = 5$ rad/s, $\eta = 0.5$, $\alpha = 0.8$, and $r = 1$

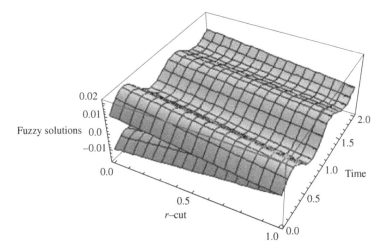

Figure 8.19 Fuzzy unit impulse response for $\omega = 5$ rad/s, $\eta = 0.5$, and $\alpha = 0.2$ (Case 1)

Case 1 $\omega = 5$ rad/s, $\eta = 0.5$ and $\alpha = 0.2$.

Case 2 $\omega = 10$ rad/s, $\eta = 0.5$, and $\alpha = 0.5$.

Case 3 $\omega = 5$ rad/s, $\eta = 0.5$, and $\alpha = 0.8$.

Case 4 $\omega = 10$ rad/s, $\eta = 0.5$, and $\alpha = 0.2$.

Accordingly, for all the cases from first to four, the obtained fuzzy unit impulse responses are shown in Figs. 8.19–8.22, respectively. Similar interpretations may be drawn as mentioned in the problem of fuzzy step responses about change of parametric values and the corresponding results.

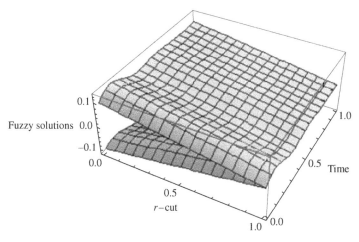

Figure 8.20 Fuzzy unit impulse response for $\omega = 10$ rad/s, $\eta = 0.5$, and $\alpha = 0.5$ (Case 2)

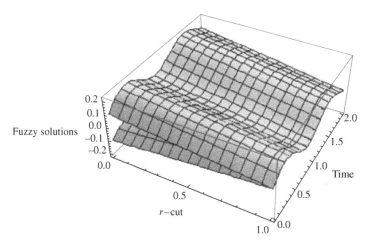

Figure 8.21 Fuzzy unit impulse response for $\omega = 5$ rad/s, $\eta = 0.05$, and $\alpha = 0.8$ (Case 3)

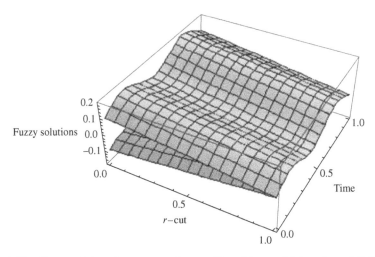

Figure 8.22   Fuzzy unit impulse response for $\omega = 10$ rad/s, $\eta = 0.05$, and $\alpha = 0.2$ (Case 4)

Here, we have included few representative results with respect to various parameters involved. For both the problems for $r = 1$, fuzzy initial conditions convert into crisp initial conditions. It is interesting to note that for both the responses (unit step and impulse), the lower and upper bounds of the fuzzy solutions are the same for $r = 1$, which are the results obtained in Behera and Chakraverty (2013).

HPM has also successfully been applied to obtain the uncertain dynamic responses of fuzzy fractionally damped simply supported beam using double parametric form of fuzzy numbers. Double parametric form of fuzzy numbers converts the corresponding differential equation in crisp form, which is found to be efficient and straightforward to solve.

BIBLIOGRAPHY

Agrawal RP, Lakshmikantham V, Nieto JJ. On the concept of solution for fractional differential equations with uncertainty. *Nonlinear Anal* 2010;**72**:2859–2862.

Arshad S, Lupulescu V. On the fractional differential equations with uncertainty. *Nonlinear Anal* 2011;**74**:3685–3693.

Behera D, Chakraverty S. Numerical solution of fractionally damped beam by homotopy perturbation method. *Cent Eur J Phys* 2013;**11**:792–798.

Chakraverty S, Behera D. Dynamic responses of fractionally damped mechanical system using homotopy perturbation method. *Alexandria Eng J* 2013;**52**:557–562.

He JH. Homotopy perturbation technique. *Comput Methods Appl Mech Eng* 1999;**178**: 257–262.

He JH. A coupling method of a homotopy technique and a perturbation technique for non-linear problems. *Int J Nonlinear Mech* 2000;**35**:37–43.

Kilbas AA, Srivastava HM, Trujillo JJ. *Theory and Applications of Fractional Differential Equations*. New York: Elsevier; 2006.

Kiryakova VS. *Generalized Fractional Calculus and Applications*. England: Longman Scientific and Technical; 1993.

Miller KS, Ross B. *An Introduction to the Fractional Calculus and Fractional Differential Equations*. New York: John Wiley & Sons; 1993.

Oldham KB, Spanier J. *The Fractional Calculus*. New York, NY: Academic Press; 1974.

Podlubny I. *Fractional Differential Equations*. New York, NY: Academic Press; 1999.

Ross TJ. *Fuzzy Logic with Engineering Applications*. New York: John Wiley & Sons; 2004.

Samko SG, Kilbas AA, Marichev OI. *Fractional Integrals and Derivatives: Theory and Application*. Langhorne, PA: Gordon and Breach Science Publishers; 1993.

Suarez LE, Shokooh A. An eigenvector expansion method for the solution of motion containing fractional derivatives. *ASME J Appl Mech* 1997;**64**:629–635.

Yuan L, Agrawal OP. A numerical scheme for dynamic systems containing fractional derivatives. *J Vib Acoust* 2002;**124**:321–324.

Zimmermann HJ. *Fuzzy Set Theory and its Application*. London: Kluwer Academic Publishers; 2001.

Zu-feng, L., and Xiao-yan, T. (2007) Analytical solution of fractionally damped beam by Adomian decomposition method," *Appl Math and Mech*, **28**, 219–227.

9

FUZZY FRACTIONAL DIFFUSION PROBLEMS

This chapter implemented the proposed double parametric form of fuzzy numbers to solve fuzzy fractional diffusion equation with initial conditions as triangular and Gaussian fuzzy number. In the solution process, homotopy perturbation method (HPM) and Adomian decomposition method (ADM) are used.

9.1 FUZZY FRACTIONAL-ORDER DIFFUSION EQUATION

First, fuzzy fractional diffusion equation has been converted to an interval-based fuzzy fractional diffusion equation using a single parametric form of fuzzy numbers. Then the interval-based differential equation is reduced to a crisp differential equation by using double parametric form of fuzzy numbers. Then HPM and ADM have been applied to obtain the solution in double parametric form.

Let us now consider the following fuzzy fractional diffusion equation:

$$\frac{\partial^\alpha \widetilde{v}(x,t)}{\partial t^\alpha} = K \frac{\partial^2 \widetilde{v}(x,t)}{\partial t^2} - \frac{\partial}{\partial x}\left(F(x)\widetilde{v}(x,t)\right), \quad 0 < \alpha \leq 1, \ t > 0, \ x > 0 \qquad (9.1)$$

with fuzzy initial condition $\widetilde{v}(x,0) = \widetilde{f}(x)$, where $\partial^\alpha/\partial t^\alpha$ is the Caputo derivative of order α, $\widetilde{v}(x,t)$ represents the uncertain probability density function of finding a particle at x in time t, K is the constant that depends on the temperature, and $F(x)$

Fuzzy Arbitrary Order System: Fuzzy Fractional Differential Equations and Applications, First Edition.
Snehashish Chakraverty, Smita Tapaswini, and Diptiranjan Behera.
© 2016 John Wiley & Sons, Inc. Published 2016 by John Wiley & Sons, Inc.

is the external force. In the present analysis, we have considered $F(x) = -g(x)$. Equation (9.1) may be written as

$$\frac{\partial^\alpha \widetilde{v}(x,t)}{\partial t^\alpha} = K\frac{\partial^2 \widetilde{v}(x,t)}{\partial x^2} + \frac{\partial}{\partial x}\left(g(x)\widetilde{v}(x,t)\right), \qquad (9.2)$$

with fuzzy initial condition $\widetilde{v}(x,0) = \widetilde{f}(x)$.

The aforementioned fuzzy fractional differential equation (Eq. (9.2)) can be written as

$$\left[\frac{\partial^\alpha \underline{v}(x,t;r)}{\partial t^\alpha}, \frac{\partial^\alpha \overline{v}(x,t;r)}{\partial t^\alpha}\right] = K\left[\frac{\partial^2 \underline{v}(x,t;r)}{\partial x^2}, \frac{\partial^2 \overline{v}(x,t;r)}{\partial x^2}\right]$$
$$+ g(x)\left[\frac{\partial \underline{v}(x,t;r)}{\partial x}, \frac{\partial \overline{v}(x,t;r)}{\partial x}\right], \qquad (9.3)$$

subject to fuzzy initial condition

$$\left[\underline{v}(x,0;r), \overline{v}(x,0;r)\right] = \left[\underline{f}(x;r), \overline{f}(x;r)\right].$$

Using the double parametric form (as discussed in Chapter 1), Eq. (9.3) can be expressed as

$$\left\{\beta\left(\frac{\partial^\alpha \overline{v}(x,t;r)}{\partial t^\alpha} - \frac{\partial^\alpha \underline{v}(x,t;r)}{\partial t^\alpha}\right) + \frac{\partial^\alpha \underline{v}(x,t;r)}{\partial t^2}\right\}$$
$$= \left\{\beta\left(\frac{\partial^2 \overline{v}(x,t;r)}{\partial x^2} - \frac{\partial^2 \underline{v}(x,t;r)}{\partial x^2}\right) + \frac{\partial^2 \underline{v}(x,t;r)}{\partial x^2}\right\}$$
$$+ g(x)\left\{\beta\left(\frac{\partial \overline{v}(x,t;r)}{\partial x} - \frac{\partial \underline{v}(x,t;r)}{\partial x}\right) + \frac{\partial \underline{v}(x,t;r)}{\partial x}\right\}, \qquad (9.4)$$

subject to the fuzzy initial condition

$$\left\{\beta(\overline{v}(x,0;r) - \underline{v}(x,0;r)) + \underline{v}(x,0;r)\right\} = \beta\left(\overline{f}(x;r) - \underline{f}(x;r)\right) + \underline{f}(x;r),$$

$$\text{where, } r, \beta \in [0,1].$$

Let us now denote

$$\left\{\beta\left(\frac{\partial^\alpha \overline{v}(x,t;r)}{\partial t^\alpha} - \frac{\partial^\alpha \underline{v}(x,t;r)}{\partial t^\alpha}\right) + \frac{\partial^\alpha \underline{v}(x,t;r)}{\partial t^2}\right\} = \frac{\partial^\alpha \widetilde{v}(x,t;r,\beta)}{\partial t^\alpha},$$

$$\left\{\beta\left(\frac{\partial^2 \overline{v}(x,t;r)}{\partial x^2} - \frac{\partial^2 \underline{v}(x,t;r)}{\partial x^2}\right) + \frac{\partial^2 \underline{v}(x,t;r)}{\partial x^2}\right\} = \frac{\partial^2 \widetilde{v}(x,t;r,\beta)}{\partial x^2},$$

$$\left\{\beta\left(\frac{\partial \overline{v}(x,t;r)}{\partial x} - \frac{\partial \underline{v}(x,t;r)}{\partial x}\right) + \frac{\partial \underline{v}(x,t;r)}{\partial x}\right\} = \frac{\partial \widetilde{v}(x,t;r,\beta)}{\partial x},$$

$$\left\{ \beta \left(\overline{v}(x,0;r) - \underline{v}(x,0;r) \right) + \underline{v}(x,0;r) \right\} = \widetilde{v}(x,0;r,\beta) \text{ and}$$

$$\beta \left(\overline{f}(x;r) - \underline{f}(x;r) \right) + \underline{f}(x;r) = \widetilde{f}(x;r;\beta).$$

Substituting these in Eq. (9.4), we get

$$\frac{\partial^{\alpha} \widetilde{v}(x,t;r,\beta)}{\partial t^{\alpha}} = K \frac{\partial^2 \widetilde{v}(x,t;r,\beta)}{\partial x^2} + \frac{\partial}{\partial x} \left(g(x) \widetilde{v}(x,t;r,\beta) \right), \tag{9.5}$$

with initial condition

$$\widetilde{v}(x,0;r,\beta) = \widetilde{f}(x;r;\beta).$$

Solving the corresponding crisp differential equation (Eq. (9.5)), one may get the solution as $\widetilde{v}(x,t;r,\beta)$. The lower and upper bounds of the solution in single parametric form have been obtained by substituting $\beta = 0$ and 1, respectively, which may be represented as

$$\widetilde{v}(x,t;r,0) = \underline{v}(x,t,r) \text{ and } \widetilde{v}(x,t,r,1) = \overline{v}(x,t,r)$$

Now HPM and ADM have been applied to solve Eq. (9.5).

9.1.1 Double-Parametric-Based Solution of Uncertain Fractional-Order Diffusion Equation

- Solution by HPM
    In this section, HPM has been applied to solve Eq. (9.5). A simple homotopy is constructed for an embedding parameter $p \in [0,1]$, as follows:

$$(1-p)\frac{\partial^{\alpha} \widetilde{v}(x,t;r,\beta)}{\partial t^{\alpha}} + p \left[\frac{\partial^{\alpha} \widetilde{v}(x,t;r,\beta)}{\partial t^{\alpha}} - K \frac{\partial^2 \widetilde{v}(x,t;r,\beta)}{\partial x^2} - g(x)\frac{\partial \widetilde{v}(x,t;r,\beta)}{\partial x} \right]$$
$$= 0 \tag{9.6}$$

or

$$\frac{\partial^{\alpha} \widetilde{v}(x,t;r,\beta)}{\partial t^{\alpha}} + p \left[-K \frac{\partial^2 \widetilde{v}(x,t;r,\beta)}{\partial x^2} - g(x)\frac{\partial \widetilde{v}(x,t;r,\beta)}{\partial x} \right] = 0. \tag{9.7}$$

In the changing process from 0 to 1, for $p = 0$, Eq. (9.6) or (9.7) gives $\partial^{\alpha} \widetilde{v}(x,t;r,\beta)/\partial t^{\alpha} = 0$ and for $p = 1$, one may have the original system

$$\frac{\partial^{\alpha} \widetilde{v}(x,t;r,\beta)}{\partial t^{\alpha}} - K \frac{\partial^2 \widetilde{v}(x,t;r,\beta)}{\partial x^2} - g(x)\frac{\partial \widetilde{v}(x,t;r,\beta)}{\partial x} = 0.$$

This is called deformation in topology.

$$\frac{\partial^{\alpha} \widetilde{v}(x,t;r,\beta)}{\partial t^{\alpha}} \text{ and } -K \frac{\partial^2 \widetilde{v}(x,t;r,\beta)}{\partial x^2} - g(x)\frac{\partial \widetilde{v}(x,t;r,\beta)}{\partial x}$$

are called homotopic. We can assume the solution of Eq. (9.1) or (9.2) as a power series expansion in p as

$$\widetilde{v}(x,t;r,\beta) = \widetilde{v}_0(x,t;r,\beta) + p\widetilde{v}_1(x,t;r,\beta) + p^2\widetilde{v}_2(x,t;r,\beta)$$
$$+ p^3\widetilde{v}_3(x,t;r,\beta) + \cdots, \tag{9.8}$$

where, $v_i(x,t;r,\beta)$ for $i = 0, 1, 2, 3, \ldots$ are functions yet to be determined. Substituting Eq. (9.8) into Eq. (9.6) or (9.7) and equating the terms with the identical powers of p, we have

$$p^0 : \frac{\partial^\alpha \widetilde{v}_0(x,t;r,\beta)}{\partial t^\alpha} = 0, \tag{9.9}$$

$$p^1 : \frac{\partial^\alpha \widetilde{v}_1(x,t;r,\beta)}{\partial t^\alpha} - \frac{\partial^2 \widetilde{v}_0(x,t;r,\beta)}{\partial x^2} - g(x)\frac{\partial \widetilde{v}_0(x,t;r,\beta)}{\partial x} = 0, \tag{9.10}$$

$$p^2 : \frac{\partial^\alpha \widetilde{v}_2(x,t;r,\beta)}{\partial t^\alpha} - \frac{\partial^2 \widetilde{v}_1(x,t;r,\beta)}{\partial x^2} - g(x)\frac{\partial \widetilde{v}_1(x,t;r,\beta)}{\partial x} = 0, \tag{9.11}$$

$$p^3 : \frac{\partial^\alpha \widetilde{v}_3(x,t;r,\beta)}{\partial t^\alpha} - \frac{\partial^2 \widetilde{v}_2(x,t;r,\beta)}{\partial x^2} - g(x)\frac{\partial \widetilde{v}_2(x,t;r,\beta)}{\partial x} = 0, \tag{9.12}$$

$$p^4 : \frac{\partial^\alpha \widetilde{v}_4(x,t;r,\beta)}{\partial t^\alpha} - \frac{\partial^2 \widetilde{v}_3(x,t;r,\beta)}{\partial x^2} - g(x)\frac{\partial \widetilde{v}_3(x,t;r,\beta)}{\partial x} = 0, \tag{9.13}$$

and so on.

Choosing initial approximation $\widetilde{v}(x,0;r,\beta)$ and applying the operator J^α (the inverse operator of Caputo derivative of order α) on both sides of Eqs. (9.9)–(9.13), one may get the approximate solution $\widetilde{v}(x,t,r,\beta) = \lim_{p\to 1} \widetilde{v}(x,t;r,\beta)$, which can be expressed as

$$\widetilde{v}(x,t;r,\beta) = \widetilde{v}_0(x,t;r,\beta) + \widetilde{v}_1(x,t;r,\beta) + \widetilde{v}_2(x,t;r,\beta) + \widetilde{v}_3(x,t;r,\beta) + \cdots. \tag{9.14}$$

- Solution by ADM

  Let us consider Eq. (9.5) as

$$L_t^\alpha \widetilde{v}(x,t;r,\beta) = K L_{xx}\widetilde{v}(x,t;r,\beta) + L_x \left(g(x)\widetilde{v}(x,t;r,\beta)\right), \tag{9.15}$$

where, $L_t^\alpha \equiv \partial^\alpha/\partial t^\alpha$, $L_{xx} \equiv \partial^2/\partial x^2$, and $L_x \equiv \partial/\partial x$.

Applying the operator $L_t^{-\alpha}$ (which is the inverse operator of L_t^α) on both sides of Eq. (9.15), obtained equivalent expression is given as

$$L_t^{-\alpha} L_t^\alpha \widetilde{v}(x,t;r,\beta) = L_t^{-\alpha}\left(K L_{xx}\widetilde{v}(x,t;r,\beta) + L_x\left(g(x)\widetilde{v}(x,t;r,\beta)\right)\right). \tag{9.16}$$

Now, we have

$$L_t^{-\alpha} L_t^\alpha \widetilde{v}(x,t;r,\beta) = v(x,0;r,\beta).$$

Then Eq. (9.16) becomes

$$\widetilde{v}(x,t;r,\beta) = v(x,0;r,\beta) + L_t^{-\alpha}\left(KL_{xx}\widetilde{v}(x,t;r,\beta) + L_x\left(g(x)\widetilde{v}(x,t;r,\beta)\right)\right).$$
(9.17)

According to Adomian decomposition method (Adomian, 1984, 1994), we assume an infinite series solution for unknown function $v(x,t;r,\beta)$ as

$$v(x,t;r,\beta) = \sum_{n=0}^{\infty} v_n(x,t;r,\beta),$$
(9.18)

with $v_0(x,t;r,\beta) = v(x,0;r,\beta)$ and the components $v_n(x,t;r,\beta)$ where, $n > 0$ are usually determined by

$$v_1(x,t;r,\beta) = L_t^{-\alpha}\left(\frac{\partial^2 v_0(x,t;r,\beta)}{\partial x^2} + \frac{\partial(g(x)v_0(x,t;r,\beta))}{\partial x}\right),$$

$$v_2(x,t;r,\beta) = L_t^{-\alpha}\left(\frac{\partial^2 v_1(x,t;r,\beta)}{\partial x^2} + \frac{\partial(g(x)v_1(x,t;r,\beta))}{\partial x}\right),$$

$$v_3(x,t;r,\beta) = L_t^{-\alpha}\left(\frac{\partial^2 v_2(x,t;r,\beta)}{\partial x^2} + \frac{\partial(g(x)v_2(x,t;r,\beta))}{\partial x}\right),$$

$$v_4(x,t;r,\beta) = L_t^{-\alpha}\left(\frac{\partial^2 v_3(x,t;r,\beta)}{\partial x^2} + \frac{\partial(g(x)v_3(x,t;r,\beta))}{\partial x}\right),$$

and so on.

Now substituting these terms in Eq. (9.18), one may get the approximate solution of Eq. (9.5) as follows:

$$v(x,t;r,\beta) = v_0(x,t;r,\beta) + v_1(x,t;r,\beta) + v_2(x,t;r,\beta) + v_3(x,t;r,\beta) + \cdots.$$
(9.19)

As mentioned earlier, the series obtained by ADM converges very rapidly and only few terms are required to get the approximate solutions. The proof may be found in Abbaoui and Cherruault (1994, 1995), Cherruault (1989), and Himoun, (1999).

9.1.2 Solution Bounds for Different External Forces

- Using HPM

Case 1 Let us consider $K = 1$ and $F(x) = -g(x)$ where $g(x) = x$ with fuzzy initial condition in parametric form as $\widetilde{v}(x,0;r) = [0.1r + 0.9, 1.1 - 0.1r]$. Hence, Eq. (9.2) will become

$$\frac{\partial^\alpha \widetilde{v}(x,t)}{\partial t^\alpha} = K\frac{\partial^2 \widetilde{v}(x,t)}{\partial x^2} + \frac{\partial}{\partial x}\left(x\widetilde{v}(x,t)\right).$$
(9.20)

Using double parametric form, Eq. (9.20) and the corresponding fuzzy initial condition will become

$$\frac{\partial^\alpha \widetilde{v}(x,t;r,\beta)}{\partial t^\alpha} = K\frac{\partial^2 \widetilde{v}(x,t;r,\beta)}{\partial x^2} + \frac{\partial}{\partial x}\left(x\widetilde{v}(x,t,r,\beta)\right)$$

and

$$\widetilde{v}(x,0;r,\beta) = \beta(0.2 - 0.2r) + (0.1r + 0.9).$$

Applying HPM, we have

$$\widetilde{v}_0(x,t;r,\beta) = \{\beta(0.2 - 0.2r) + (0.1r + 0.9)\}, \tag{9.21}$$

$$\widetilde{v}_1(x,t;r,\beta) = \{\beta(0.2 - 0.2r) + (0.1r + 0.9)\}\frac{t^\alpha}{\Gamma(\alpha + 1)}, \tag{9.22}$$

$$\widetilde{v}_2(x,t;r,\beta) = \{\beta(0.2 - 0.2r) + (0.1r + 0.9)\}\frac{t^{2\alpha}}{\Gamma(2\alpha + 1)}, \tag{9.23}$$

$$\widetilde{v}_3(x,t;r,\beta) = \{\beta(0.2 - 0.2r) + (0.1r + 0.9)\}\frac{t^{3\alpha}}{\Gamma(3\alpha + 1)}, \tag{9.24}$$

$$\widetilde{v}_4(x,t;r,\beta) = \{\beta(0.2 - 0.2r) + (0.1r + 0.9)\}\frac{t^{4\alpha}}{\Gamma(4\alpha + 1)}, \tag{9.25}$$

and so on.

Therefore, the solution can be written in general form as

$$\widetilde{v}(x,t;r,\beta) = \{\beta(0.2 - 0.2r) + (0.1r + 0.9)\}\sum_{n=0}^{\infty}\frac{t^{n\alpha}}{\Gamma(n\alpha + 1)}. \tag{9.26}$$

To obtain the solution bound in single parametric form, we may substitute $\beta = 0$ and 1 for lower and upper bounds of the solution, respectively, to get

$$\underline{v}(x,t;r,0) = \underline{v}(x,t;r) = (0.1r + 0.9)\sum_{n=0}^{\infty}\frac{t^{n\alpha}}{\Gamma(n\alpha + 1)} \tag{9.27}$$

and

$$\overline{v}(x,t;r,1) = \overline{v}(t;r,1) = (1.1 - 0.1r)\sum_{n=0}^{\infty}\frac{t^{n\alpha}}{\Gamma(n\alpha + 1)}. \tag{9.28}$$

We note that in the special case when $r = 1$, the results (crisp) obtained by the proposed method are exactly same as those obtained by the method of Godal et al. (2013).

Case 2   Next, we consider the same force and constant K as in Case 1 with different fuzzy initial condition, namely $\widetilde{v}(x,0;r) = x\,[0.1r + 0.9, 1.1 - 0.1r]$.

Again, by applying the procedure discussed previously, we get the solution

$$\widetilde{v}_0(x, t; r, \beta) = x\left\{\beta(0.2 - 0.2r) + (0.1r + 0.9)\right\}, \tag{9.29}$$

$$\widetilde{v}_1(x, t; r, \beta) = 2x\left\{\beta(0.2 - 0.2r) + (0.1r + 0.9)\right\}\frac{t^\alpha}{\Gamma(\alpha + 1)}, \tag{9.30}$$

$$\widetilde{v}_2(x, t; r, \beta) = 4x\left\{\beta(0.2 - 0.2r) + (0.1r + 0.9)\right\}\frac{t^{2\alpha}}{\Gamma(2\alpha + 1)}, \tag{9.31}$$

$$\widetilde{v}_3(x, t; r, \beta) = 8x\left\{\beta(0.2 - 0.2r) + (0.1r + 0.9)\right\}\frac{t^{3\alpha}}{\Gamma(3\alpha + 1)}, \tag{9.32}$$

$$\widetilde{v}_4(x, t; r, \beta) = 16x\left\{\beta(0.2 - 0.2r) + (0.1r + 0.9)\right\}\frac{t^{4\alpha}}{\Gamma(4\alpha + 1)}, \tag{9.33}$$

and so on. Substituting Eqs. (9.29)–(9.33) in Eq. (9.14), one may get the general solution of $\widetilde{v}(x, t; r, \beta)$ as

$$\widetilde{v}(x, t; r, \beta) = x\left\{\beta(0.2 - 0.2r) + (0.1r + 0.9)\right\}\sum_{n=0}^{\infty}\frac{t^{n\alpha}}{\Gamma(n\alpha + 1)}. \tag{9.34}$$

Substituting $\beta = 0$ and 1 in $\widetilde{v}(x, t; r, \beta)$, we get the lower and upper bounds of the fuzzy solutions, respectively, as

$$\underline{v}(x, t; r, 0) = \underline{v}(x, t; r) = x\,(0.1r + 0.9)\sum_{n=0}^{\infty}\frac{2^n t^{n\alpha}}{\Gamma(n\alpha + 1)} \tag{9.35}$$

and

$$\overline{v}(x, t; r, 1) = \overline{v}(x, t; r) = x\,(1.1 - 0.1r)\sum_{n=0}^{\infty}\frac{2^n t^{n\alpha}}{\Gamma(n\alpha + 1)}. \tag{9.36}$$

The solution obtained by proposed method for $r = 1$ (crisp result) is again found to be exactly same as that of Godal et al. (2013).

Case 3   Finally, we consider the case when $K = 1$ and $F(x) = -g(x)$ where $g(x) = e^{-x}$ with fuzzy initial condition in parametric form as $\widetilde{v}(x, 0; r) = e^x[0.1r + 0.9, 1.1 - 0.1r]$.

Similarly, we have the following fuzzy fractional diffusion equation as

$$\frac{\partial^\alpha \widetilde{v}(x, t)}{\partial t^\alpha} = K\frac{\partial^2 \widetilde{v}(x, t)}{\partial x^2} + \frac{\partial}{\partial x}\left(e^{-x}\widetilde{v}(x, t)\right). \tag{9.37}$$

By following the proposed method with HPM, we get the solution in double parametric form as

$$\widetilde{v}(x, t; r, \beta) = e^x\left\{\beta(0.2 - 0.2r) + (0.1r + 0.9)\right\}\sum_{n=0}^{\infty}\frac{t^{n\alpha}}{\Gamma(n\alpha + 1)}. \tag{9.38}$$

The lower and upper bounds of the fuzzy solution of Eq. (9.37) are obtained as

$$\widetilde{v}(x, t; r, 0) = e^x(0.1r + 0.9) \sum_{n=0}^{\infty} \frac{t^{n\alpha}}{\Gamma(n\alpha + 1)}$$

and

$$\widetilde{v}(x, t; r, 1) = e^x(1.1 - 0.1r) \sum_{n=0}^{\infty} \frac{t^{n\alpha}}{\Gamma(n\alpha + 1)}.$$

Obtained results for $r = 1$ exactly agree with the exact solution (crisp) of Godal et al. (2013).

- Using ADM

    In this section, uncertain solution bounds for Eq. (9.5) using ADM with double parametric form of fuzzy numbers have been discussed. Depending upon K, $F(x)$, and $g(x)$, different cases (Cases 4–6) similar to Godal et al. (2013) have been analyzed in the following paragraphs.

Case 4 **(Same as Case 1 of HPM)** Here $K = 1$ and $F(x) = -g(x)$ where $g(x) = x$ with fuzzy initial condition in parametric form as $\widetilde{v}(x, 0; r) = [0.1r + 0.9, 1.1 - 0.1r]$ have been considered. Hence, Eq. (9.2) will become

$$\frac{\partial^\alpha \widetilde{v}(x, t)}{\partial t^\alpha} = K \frac{\partial^2 \widetilde{v}(x, t)}{\partial x^2} + \frac{\partial}{\partial x}(x\widetilde{v}(x, t)), \tag{9.39}$$

with fuzzy initial condition

$$\widetilde{v}(x, 0; r) = [0.1r + 0.9, 1.1 - 0.1r].$$

Using double parametric form, Eq. (9.5) and the fuzzy initial condition become

$$\frac{\partial^\alpha \widetilde{v}(x, t; r, \beta)}{\partial t^\alpha} = K \frac{\partial^2 \widetilde{v}(x, t; r, \beta)}{\partial x^2} + \frac{\partial}{\partial x}(x\widetilde{v}(x, t, r, \beta),$$

with the initial condition

$$\widetilde{v}(0; r, \beta) = \beta(0.2 - 0.2r) + (0.1r + 0.9).$$

By using ADM we have

$$\widetilde{v}_0(x, t; r, \beta) = \{\beta(0.2 - 0.2r) + (0.1r + 0.9)\}, \tag{9.40}$$

$$\widetilde{v}_1(x, t; r, \beta) = \{\beta(0.2 - 0.2r) + (0.1r + 0.9)\} \frac{t^\alpha}{\Gamma(\alpha + 1)}, \tag{9.41}$$

$$\widetilde{v}_2(x, t; r, \beta) = \{\beta(0.2 - 0.2r) + (0.1r + 0.9)\} \frac{t^{2\alpha}}{\Gamma(2\alpha + 1)}, \tag{9.42}$$

$$\tilde{v}_3(x, t; r, \beta) = \{\beta(0.2 - 0.2r) + (0.1r + 0.9)\} \frac{t^{3\alpha}}{\Gamma(3\alpha + 1)}, \qquad (9.43)$$

$$\tilde{v}_4(x, t; r, \beta) = \{\beta(0.2 - 0.2r) + (0.1r + 0.9)\} \frac{t^{4\alpha}}{\Gamma(4\alpha + 1)}, \qquad (9.44)$$

and so on.

Therefore, the solution can be written in general form as

$$\tilde{v}(x, t; r, \beta) = \{\beta(0.2 - 0.2r) + (0.1r + 0.9)\} \sum_{n=0}^{\infty} \frac{t^{n\alpha}}{\Gamma(n\alpha + 1)}. \qquad (9.45)$$

To obtain the solution bound in single parametric form, we may substitute $\beta = 0$ and 1 to get the lower and upper bound of the solution, respectively, as

$$\underline{v}(x, t; r, 0) = \underline{v}(x, t; r) = (0.1r + 0.9) \sum_{n=0}^{\infty} \frac{t^{n\alpha}}{\Gamma(n\alpha + 1)} \qquad (9.46)$$

and

$$\overline{v}(x, t; r, 1) = \overline{v}(t; r, 1) = (1.1 - 0.1r) \sum_{n=0}^{\infty} \frac{t^{n\alpha}}{\Gamma(n\alpha + 1)}. \qquad (9.47)$$

One may note that in the special case when $r = 1$, the results (crisp) obtained by the proposed method are exactly same as those obtained by the method of Godal et al. (2013).

Case 5 **(Same as Case 2 of HPM)** The same force and constant K as Case 1 have been assumed with different fuzzy initial condition $\tilde{v}(x, 0; r) = x[0.1r + 0.9, 1.1 - 0.1r]$.

Again, by applying the procedure discussed previously, we get the solution

$$\tilde{v}_0(x, t; r, \beta) = x\{\beta(0.2 - 0.2r) + (0.1r + 0.9)\}, \qquad (9.48)$$

$$\tilde{v}_1(x, t; r, \beta) = 2x\{\beta(0.2 - 0.2r) + (0.1r + 0.9)\} \frac{t^{\alpha}}{\Gamma(\alpha + 1)}, \qquad (9.49)$$

$$\tilde{v}_2(x, t; r, \beta) = 4x\{\beta(0.2 - 0.2r) + (0.1r + 0.9)\} \frac{t^{2\alpha}}{\Gamma(2\alpha + 1)}, \qquad (9.50)$$

$$\tilde{v}_3(x, t; r, \beta) = 8x\{\beta(0.2 - 0.2r) + (0.1r + 0.9)\} \frac{t^{3\alpha}}{\Gamma(3\alpha + 1)}, \qquad (9.51)$$

$$\tilde{v}_4(x, t; r, \beta) = 16x\{\beta(0.2 - 0.2r) + (0.1r + 0.9)\} \frac{t^{4\alpha}}{\Gamma(4\alpha + 1)}, \qquad (9.52)$$

and so on. Substituting Eqs. (9.48)–(9.52) in Eq. (9.19), one may get the general solution of $\tilde{v}(x, t; r, \beta)$ as

$$\tilde{v}(x, t; r, \beta) = x\{\beta(0.2 - 0.2r) + (0.1r + 0.9)\} \sum_{n=0}^{\infty} \frac{t^{n\alpha}}{\Gamma(n\alpha + 1)}. \qquad (9.53)$$

Substituting $\beta = 0$ and 1 in $\tilde{v}(x, t; r, \beta)$, the lower and upper bounds of the fuzzy solutions have been obtained, respectively, as

$$\underline{v}(x, t; r, 0) = \underline{v}(x, t; r) = x\,(0.1r + 0.9) \sum_{n=0}^{\infty} \frac{2^n t^{n\alpha}}{\Gamma(n\alpha + 1)} \tag{9.54}$$

and

$$\overline{v}(x, t; r, 1) = \overline{v}(x, t; r) = x\,(1.1 - 0.1r) \sum_{n=0}^{\infty} \frac{2^n t^{n\alpha}}{\Gamma(n\alpha + 1)}. \tag{9.55}$$

The solution obtained by the proposed method for $r = 1$ is again found to be exactly the same as that of (crisp result) Godal et al. (2013).

Case 6   Next, we consider $K = 1$ and $F(x) = -g(x)$ where $g(x) = e^{-x}$ with fuzzy initial condition in parametric form as $\tilde{v}(x, 0; r) = e^x[1 - \sqrt{-0.18\log_e r}, 1 + \sqrt{-0.18\log_e r}]$.

We have the following fuzzy fractional diffusion equation as

$$\frac{\partial^{\alpha} \tilde{v}(x, t)}{\partial t^{\alpha}} = K \frac{\partial^2 \tilde{v}(x, t)}{\partial x^2} + \frac{\partial}{\partial x} \left(e^{-x} \tilde{v}(x, t) \right). \tag{9.56}$$

By ADM, the solution in double parametric form may be written as

$$\tilde{v}(x, t; r, \beta) = e^x \left\{ \beta \left(2\sqrt{-0.18\log_e r} \right) + \left(1 - \sqrt{-0.18\log_e r} \right) \right\} \sum_{n=0}^{\infty} \frac{t^{n\alpha}}{\Gamma(n\alpha + 1)}. \tag{9.57}$$

The lower and upper bounds of the fuzzy solution of Eq. (9.56) are obtained as

$$\tilde{v}(x, t; r, 0) = e^x \left(1 - \sqrt{-0.18\log_e r} \right) \sum_{n=0}^{\infty} \frac{t^{n\alpha}}{\Gamma(n\alpha + 1)}$$

and

$$\tilde{v}(x, t; r, 1) = e^x \left(1 + \sqrt{-0.18\log_e r} \right) \sum_{n=0}^{\infty} \frac{t^{n\alpha}}{\Gamma(n\alpha + 1)}.$$

It may be noted that results for $r = 1$ exactly agree with the exact solution (crisp) of Godal et al. (2013).

9.2   NUMERICAL RESULTS OF FUZZY FRACTIONAL DIFFUSION EQUATION

Numerical results for fuzzy fractional diffusion equation with different external forces and fuzzy initial conditions are computed by the method of HPM. Obtained results by the present analysis are compared with the existing solution of Godal et al. (2013) in special cases to show the validation of the proposed analysis.

Triangular fuzzy solutions for Cases 1–3 are depicted in Figs. 9.1–9.3 by varying the time from 0 to 0.5, 0 to 10, and 0 to 5, respectively, with $\alpha = 0.6$. For Cases 2 and 3, we have considered $x = 0.5$. Next, interval solutions for $r = 0.6$ for different cases have been given in Figs. 9.4–9.6 for $x = 0.5$ and $\alpha = 0.4$, 0.8, and 1. It may be worth mentioning that for all the cases, present results exactly agree with the solution

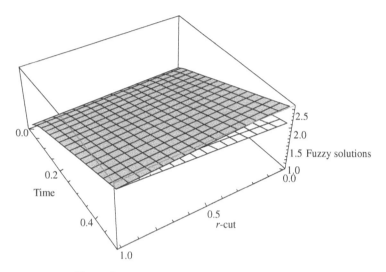

Figure 9.1 Fuzzy solution for Case 1 using HPM

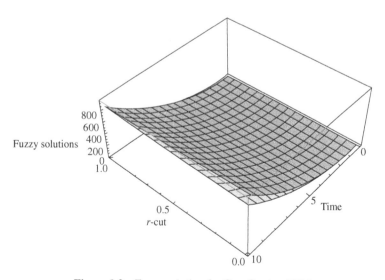

Figure 9.2 Fuzzy solution for Case 2 using HPM

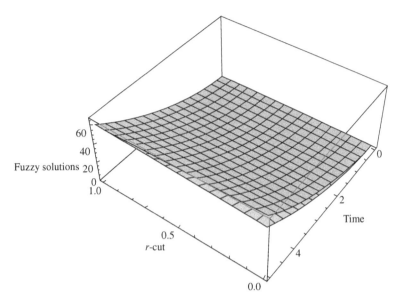

Figure 9.3   Fuzzy solution for Case 3 using HPM

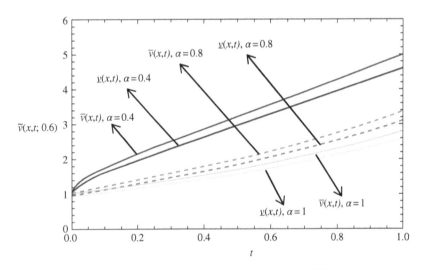

Figure 9.4   Interval solutions for Case 1 using HPM

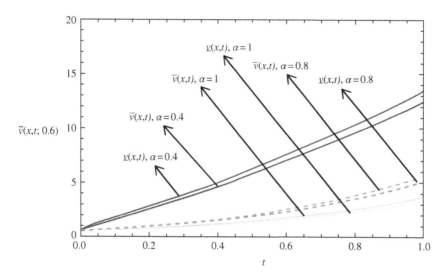

Figure 9.5   Interval solutions for Case 2 using HPM

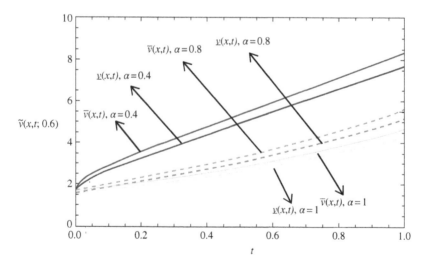

Figure 9.6   Interval solutions for Case 3 using HPM

of Godal et al. (2013) in special case of $r = 1$. Also, it is interesting to note from Figs. 9.4–9.6 that the left and right bounds of the uncertain probability density function, that is, $\tilde{v}(x, t)$ (with particular values of α, r and x) gradually decrease with the increase of the fractional order α and time t.

Computed results in digital form for the Cases 1, 2, 4, and 5 are also shown in Tables 9.1 and 9.2, respectively, by varying t from 0 to 1 and taking $x = 0.5$, $r = 0.5$ with $\alpha = 0.4$. It is interesting to note that in special case (for $r = 1$), the results exactly match with the Godal et al. (2013).

In a similar manner, the numerical results are discussed here with ADM. Triangular fuzzy solutions for Cases 4 and 5 and Gaussian fuzzy solution for Case 6 are depicted in Figs. 9.7–9.9 by varying the time from 0 to 0.5, 0 to 10, and 0 to 5, respectively, with $\alpha = 0.5$. For Cases 5 and 6, we have considered $x = 0.5$. Next, interval solutions for $r = 0.6$ and 1 for different cases have been shown in Figs. 9.10–9.15 for

TABLE 9.1 Comparison of Present with HPM, Present with ADM and Godal et al. (2013) Solutions of Cases 1 and 4 for $x = 0.5$ and $\alpha = 0.4$

t	$\underline{v}(x, t)$ $r = 0.5$		$\overline{v}(x, t)$ $r = 0.5$		$\underline{v}(x, t) = \overline{v}(x, t) = v(x, t)$ $r = 1$		$v(x, t)$
	Present with HPM	Present with ADM	Present with HPM	Present with ADM	Present with HPM	Present with ADM	Godal et al. (2013)
0	0.95	1.05	0.95	1.05	1	1	1
0.2	1.9695	2.1768	1.9695	2.1768	2.0732	2.0732	2.0732
0.4	2.6227	2.8988	2.6227	2.8988	2.7608	2.7608	2.7608
0.6	3.2612	3.6045	3.2612	3.6045	3.4328	3.4328	3.4328
0.8	3.9072	4.3185	3.9072	4.3185	4.1128	4.1128	4.1128
1	4.5674	5.0482	4.5674	5.0482	4.8078	4.8078	4.8078

TABLE 9.2 Comparison of Present with HPM, Present with ADM and Godal et al. (2013) Solutions of Cases 2 and 5 for $x = 0.5$ and $\alpha = 0.4$

t	$\underline{v}(x, t)$ $r = 0.5$		$\overline{v}(x, t)$ $r = 0.5$		$\underline{v}(x, t) = \overline{v}(x, t) = v(x, t)$ $r = 1$		$v(x, t)$
	Present with HPM	Present with ADM	Present with HPM	Present with ADM	Present with HPM	Present with ADM	Godal et al. (2013)
0	0.475	0.525	0.475	0.525	0.5	0.5	0.5
0.2	2.5051	2.7688	2.5051	2.7688	2.637	2.637	2.637
0.4	4.5729	5.0543	4.5729	5.0543	4.8136	4.8136	4.8136
0.6	6.9195	7.6479	6.9195	7.6479	7.2837	7.2837	7.2837
0.8	9.5194	10.521	9.5194	10.521	10.02	10.02	10.02
1	12.351	13.651	12.351	13.651	13.001	13.001	13.001

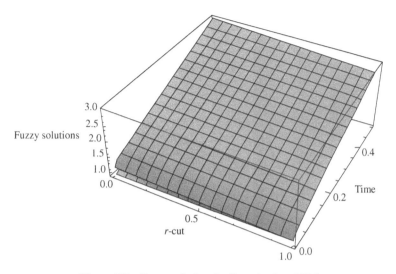

Figure 9.7   Fuzzy solution for Case 4 using ADM

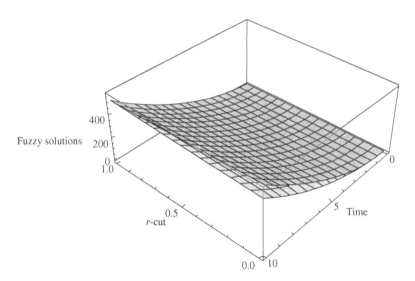

Figure 9.8   Fuzzy solution for Case 5 using ADM

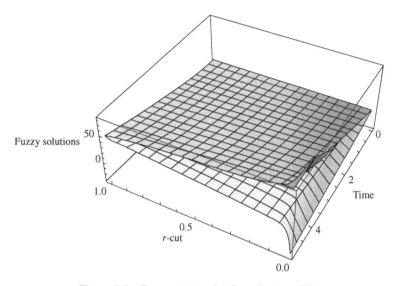

Figure 9.9   Fuzzy solution for Case 6 using ADM

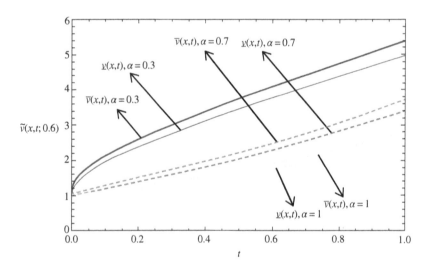

Figure 9.10   Interval solutions for $r = 0.6$ of Case 4 using ADM

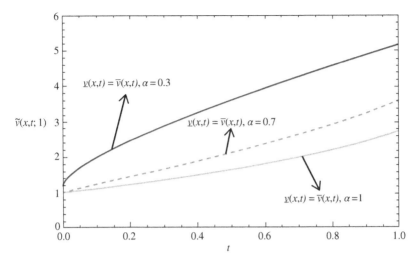

Figure 9.11   Interval solutions for $r = 1$ of Case 4 using ADM

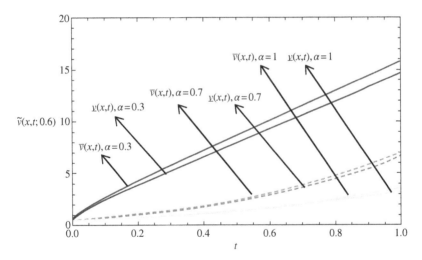

Figure 9.12   Interval solutions for $r = 0.6$ of Case 5 using ADM

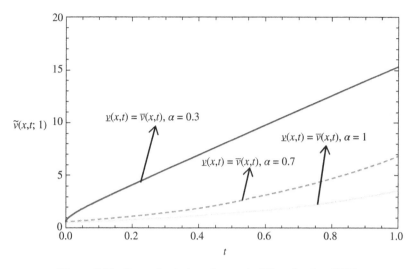

Figure 9.13   Interval solutions for $r = 1$ of Case 5 using ADM

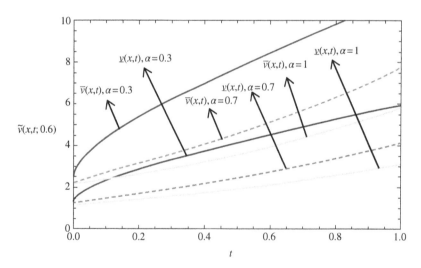

Figure 9.14   Interval solutions for $r = 0.6$ of Case 6 using ADM

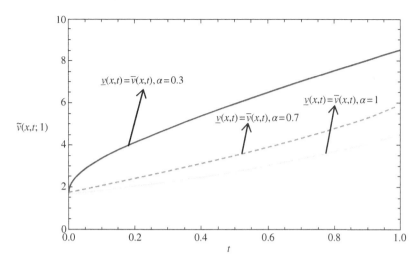

Figure 9.15 Interval solutions for $r = 1$ of Case 6 using ADM

$x = 0.5$ and $\alpha = 0.3$, 0.7, and 1. It may be worth mentioning that for all the cases, present results exactly agree with the solution of Godal et al. (2013) in special case of $r = 1$. From Figs. 9.10–9.12, one may see that the left and right bounds of the uncertain probability density function, that is, $\tilde{v}(x, t)$ (with particular values of α, r, and x) gradually decrease with the increase of the fractional order α and time t.

BIBLIOGRAPHY

Abbaoui K, Cherruault Y. Convergence of Adomian's method applied to different equations. *Comput Math Appl* 1994;**28**:103–109.

Abbaoui K, Cherruault Y. New ideas for proving convergence of decomposition methods. *Comput Math Appl* 1995;**29**:103–108.

Adomian G. A new approach to nonlinear partial differential equations. *J Math Anal Appl* 1984;**102**:420–434.

Adomian G. *Solving Frontier Problems of Physics: The Decomposition Method*. Boston: Kluwer Academic Publishers; 1994.

Cetinkaya A, Kiymaz O. The solution of the time-fractional diffusion equation by the generalized differential transform method. *Math Comput Model* 2013;**57**:2349–2354.

Cherruault Y. Convergence of Adomian's method. *Kybernetes* 1989;**18**:31–38.

Chen YM, Wu YB. Wavelet method for a class of fractional convection–diffusion equation with variable coefficients. *J Comput Sci* 2010;**1**:146–149.

Garg M, Manohar P. Numerical solution of fractional diffusion-wave equation with two space variables by matrix method. *Fract Calc Appl Anal* 2010;**13**:191–207.

Godal MA, Salah A, Khan M, Batool SI. A novel analytical solution of a fractional diffusion problem by homotopy analysis transform method. *Neural Comput Appl* 2013;**23**:1643–1647.

Gorenflo R, Luchko Y, Mainardi F. Wright functions as scale-invariant solutions of the diffusion-wave equation. *J Comput Appl Math* 2000;**118**:175–191.

He JH. Homotopy perturbation technique. *Comput Methods Appl Mech Eng* 1999;**178**: 257–262.

He JH. A coupling method of homotopy technique and perturbation technique for nonlinear problems. *Int J Non-linear Mech* 2000;**35**:37–43.

Himoun N, Abbaoui K, Cherruault Y. New results of convergence of Adomian's method. *Keber* 1999;**28**:423–429.

Lin Y, Xu C. Finite difference/spectral approximations for the time-fractional diffusion equation. *J Comput Phys* 2007;**225**:1533–1552.

Li X, Xu M, Jiang X. Homotopy perturbation method to time-fractional diffusion equation with a moving boundary condition. *Appl Math Comput* 2009;**208**:434–439.

Mainardi F. Fractional relaxation–oscillation and fractional diffusion-wave phenomena. *Chaos Soliton Fract* 1996;**7**:1461–1477.

Mainardi F, Pagnini G. The Wright functions as solutions of the time-fractional diffusion equation. *Appl Math Comput* 2003;**141**:51–62.

Merdan M. Analytical approximate solutions of fractional convection-diffusion equation with modified Riemann–Liouville derivative by means of fractional variational iteration method. *Iranian J Sci Technol A* 2013;**1**:83–92.

Momani S. General solutions for the space-and time-fractional diffusion-wave equation. *J Phys Sci* 2006;**10**:30–43.

Odibat ZM. Rectangular decomposition method for fractional diffusion-wave equations. *Appl Math Comput* 2006;**179**:92–97.

Safari M, Danesh M. Application of Adomian's decomposition method for the analytical solution of space fractional diffusion equation. *Adv Pure Math* 2011;**1**:345–350.

Tapaswini S, Chakraverty S. Non-probabilistic solutions of uncertain fractional order diffusion equations. *Fund Inform* 2014;**133**:19–34.

10

UNCERTAIN FRACTIONAL FORNBERG–WHITHAM EQUATIONS

Study of traveling-wave solutions played a very important role in many areas of physics, and it was used to study the qualitative behaviors of wave breaking. The Fornberg–Whitham equation is a type of traveling-wave solutions called kink-like wave solutions and antikink-like wave solutions. Also, it was used to study the qualitative behaviors of wave breaking. Fractional Fornberg–Whitham equation has been investigated by few authors recently. In general, the problem has been solved by considering the associated variables and parameters as crisp/exact. Similarly to the discussions in the previous chapters, we may not have these parameters in exact form, but those may be uncertain. Accordingly, fractional Fornberg–Whitham equation has been considered with uncertain but bounded form. In particular, these uncertainties are taken here as interval, which is a special case of fuzzy number. This is because for each membership value, we actually get the parametric form of interval numbers. As such, the interval fractional Fornberg–Whitham equation has been solved using variational iteration method (VIM). This method performs very well in terms of computational efficiency.

10.1 PARAMETRIC-BASED INTERVAL FRACTIONAL FORNBERG–WHITHAM EQUATION

In view of the previous discussion, let us now consider the interval fractional Fornberg–Whitham equation.

Fuzzy Arbitrary Order System: Fuzzy Fractional Differential Equations and Applications, First Edition.
Snehashish Chakraverty, Smita Tapaswini, and Diptiranjan Behera.
© 2016 John Wiley & Sons, Inc. Published 2016 by John Wiley & Sons, Inc.

$$D_t^\alpha \widetilde{u}(x,t) + D_x \widetilde{u}(x,t) + D\widetilde{u}(x,t)D_x\widetilde{u}(x,t) = D\widetilde{u}(x,t)D_{xxx}\widetilde{u}(x,t) + 3D_x\widetilde{u}(x,t)D_{xx}\widetilde{u}(x,t)$$
$$+ D_{xxt}\widetilde{u}(x,t),$$
$$t > 0, \quad 0 < \alpha \le 1, \quad x > 0. \tag{10.1}$$

with interval initial condition

$$\widetilde{u}(x,0) = \widetilde{\delta} e^{\frac{1}{2}x},$$

where $D_t^\alpha = \partial^\alpha/\partial t^\alpha$ is the Caputo derivative of order $\alpha \in (0,1]$, $\widetilde{u}(x,t)$ is the uncertain fluid velocity, t is the time, and x is the spatial coordinate.

Here the initial condition has been taken as interval. As such, this will make the governing differential equation as uncertain and the corresponding outcome or the output (result) will be in uncertain form.

The aforementioned interval fractional order Fornberg–Whitham equation (Eq. (10.1)) may be written in parametric form as

$$\left[D_t^\alpha \underline{u}(x,t;r), \; D_t^\alpha \overline{u}(x,t;r)\right] + \left[D_x \underline{u}(x,t;r), \; D_x \overline{u}(x,t;r)\right]$$
$$+ \left[\underline{u}(x,t;r)D_x\underline{u}(x,t;r), \; \overline{u}(x,t;r)D_x\overline{u}(x,t;r)\right]$$
$$= \left[\underline{u}(x,t;r)D_{xxx}\underline{u}(x,t;r), \; \overline{u}(x,t;r)D_{xxx}\overline{u}(x,t;r)\right]$$
$$+ 3\left[D_x\underline{u}(x,t;r)D_{xx}\underline{u}(x,t;r), \; D_x\overline{u}(x,t;r)D_{xx}\overline{u}(x,t;r)\right]$$
$$+ \left[D_{xxt}\underline{u}(x,t;r), D_{xxt}\overline{u}(x,t;r)\right], \tag{10.2}$$

subject to interval initial condition

$$\left[\underline{u}(x,0), \overline{u}(x,0)\right] = \left[\underline{\delta}, \overline{\delta}\right] e^{\frac{1}{2}x}.$$

Equation (10.2) may now be expressed as

$$\left\{\beta\left(D_t^\alpha \overline{u}_t^\alpha(x,t) - D_t^\alpha \underline{u}(x,t)\right) + D_t^\alpha \underline{u}(x,t)\right\}$$
$$+ \left\{\beta\left(D_x\overline{u}(x,t) - D_x\underline{u}(x,t)\right) + D_x\underline{u}(x,t)\right\}$$
$$+ \left\{\beta\left(\overline{u}(x,t)D_x\overline{u}(x,t) - \underline{u}(x,t)D_x\underline{u}(x,t)\right) + \underline{u}(x,t)D_x\underline{u}(x,t)\right\}$$
$$= \left\{\beta\left(\overline{u}(x,t)D_{xxx}\overline{u}(x,t) - \underline{u}(x,t)D_{xxx}\underline{u}(x,t)\right) + \underline{u}(x,t)D_{xxx}\underline{u}(x,t)\right\}$$
$$+ 3\left\{\beta\left(D_x\overline{u}(x,t)D_{xx}\overline{u}_{xx}(x,t) - D_x\underline{u}(x,t)D_{xx}\underline{u}(x,t)\right) + D_x\underline{u}(x,t)D_{xx}\underline{u}(x,t)\right\}$$
$$+ \left\{\beta\left(D_{xxt}\overline{u}(x,t) - D_{xxt}\underline{u}(x,t)\right) + D_{xxt}\underline{u}(x,t)\right\}, \tag{10.3}$$

subject to the initial condition

$$\left\{\beta\left(\overline{u}(x,0) - \underline{u}(x,0)\right) + \underline{u}(x,0)\right\} = \left\{\beta\left(\overline{\delta} - \underline{\delta}\right) + \underline{\delta}\right\} e^{\frac{1}{2}x}, \quad \text{where } \beta \in [0,1].$$

Let us now denote $\left\{\beta\left(D_t^\alpha\overline{u}(x,t) - D_t^\alpha\underline{u}(x,t)\right) + D_t^\alpha\underline{u}(x,t)\right\} = D_t^\alpha\widetilde{u}(x,t;\beta)$,

$\left\{\beta\left(D_x\overline{u}(x,t) - D_x\underline{u}(x,t)\right) + D_x\underline{u}(x,t)\right\} = D_x\widetilde{u}(x,t;\beta)$,

$\left\{\beta\left(\overline{u}(x,t)D_x\overline{u}(x,t) - \underline{u}(x,t)D_x\underline{u}(x,t)\right) + \underline{u}(x,t)D_x\underline{u}(x,t)\right\} = \widetilde{u}(x,t;\beta)D_x\widetilde{u}(x,t;\beta)$,

$\left\{\beta\left(\overline{u}(x,t)D_{xxx}\overline{u}(x,t) - \underline{u}(x,t)D_{xxx}\underline{u}(x,t)\right) + \underline{u}(x,t)D_{xxx}\underline{u}(x,t)\right\}$
$\qquad = \widetilde{u}(x,t;\beta)D_{xxx}\widetilde{u}(x,t;\beta)$,

$\left\{\beta\left(D_x\overline{u}(x,t)D_{xx}\overline{u}(x,t) - D_x\underline{u}(x,t)D_{xx}\underline{u}(x,t)\right) + D_x\underline{u}(x,t)D_{xx}\underline{u}(x,t)\right\}$
$\qquad = D_x\widetilde{u}(x,t;\beta)D_{xx}\widetilde{u}(x,t;\beta)$,

$\{\beta(D_{xxt}\overline{u}(x,t) - D_{xxt}\underline{u}(x,t)) + D_{xxt}\underline{u}(x,t)\} = D_{xxt}\widetilde{u}(x,t;\beta)$,

$\left\{\beta\left(\overline{u}(x,0) - \underline{u}(x,0)\right) + \underline{u}(x,0)\right\} = \widetilde{u}(x,0;\beta)$ and $\{\beta(\overline{\delta} - \underline{\delta}) + \underline{\delta}\} = \widetilde{\delta}(\beta)$.

Substituting these in Eq. (10.3), one may get

$$D_t^\alpha\widetilde{u}(x,t;\beta) + D_x\widetilde{u}(x,t;\beta) + \widetilde{u}(x,t;\beta)D_x\widetilde{u}(x,t;\beta)$$
$$= \widetilde{u}(x,t;\beta)D_{xxx}\widetilde{u}(x,t;\beta) + 3D_x\widetilde{u}(x,t;\beta)D_{xx}\widetilde{u}(x,t;\beta) + D_{xxt}\widetilde{u}(x,t;\beta), \qquad (10.4)$$

with initial condition

$$\widetilde{u}(x,0;\beta) = \widetilde{\delta}(\beta)e^{\frac{1}{2}x}. \qquad (10.5)$$

Solving the corresponding crisp differential equation (Eq. (10.4)), one may get the solution as $\widetilde{u}(x,t;\beta)$. To obtain the lower and upper bounds of the solution, we substitute $\beta = 0$ and 1, respectively, which may be represented as

$$\widetilde{u}(x,t;0) = \underline{u}(x,t) \text{ and } \widetilde{u}(x,t;1) = \overline{u}(x,t).$$

10.2   SOLUTION BY VIM (HE, 1999, 2000)

In this section, VIM has been applied to solve Eq. (10.4). According to VIM, one may construct a correction functional as follows:

$$\widetilde{u}_{n+1}(x,t;\beta) = \widetilde{u}_n(x,t;\beta) + \int_0^t \lambda(\tau)\left\{\begin{array}{l} D_t^\alpha\widetilde{u}_n(x,\tau;\beta) + D_x\widetilde{u}_n(x,\tau;\beta) \\ +\widetilde{u}_n(x,\tau;\beta)D_x\widetilde{u}_n(x,\tau;\beta) \\ -\widetilde{u}_n(x\tau;\beta)D_{xxx}\widetilde{u}_n(x,\tau;\beta) \\ -3D_x\widetilde{u}_n(x,\tau;\beta)D_{xx}\widetilde{u}_n(x,\tau;\beta) \\ -D_{xxt}\widetilde{u}_n(x,\tau;\beta) \end{array}\right\}d\tau. \qquad (10.6)$$

Making the aforementioned correction functional, that is, Eq. (10.6) stationary, and noticing that $\delta \widetilde{u}_n = 0$, we obtain

$$\delta \widetilde{u}_{n+1}(x,t;\beta) = \delta \widetilde{u}_n(x,t;\beta) + \delta \int_0^t \lambda(\tau) \left\{ \begin{array}{l} D_t^\alpha \widetilde{u}_n(x,\tau;\beta) + D_x \widetilde{u}_n(x,\tau;\beta) \\ +\widetilde{u}_n(x,\tau;\beta)D_x \widetilde{u}_n(x,\tau;\beta) \\ -\widetilde{u}_n(x\tau;\beta)D_{xxx} \widetilde{u}_n(x,\tau;\beta) \\ -3D_x \widetilde{u}_n(x,\tau;\beta)D_{xx} \widetilde{u}_n(x,\tau;\beta) \\ -D_{xxt} \widetilde{u}_n(x,\tau;\beta) \end{array} \right\} d\tau$$

$$= \delta \widetilde{u}_n(x,t;\beta) + \lambda(\tau)\delta \widetilde{u}_n(x,t;\beta) - \int_0^t \lambda'(\tau)\delta \widetilde{u}_n(x,\tau;\beta)d\tau = 0$$

$$= (1 + \lambda(\tau))\delta \widetilde{u}_n(x,t;\beta) - \int_0^t \lambda'(\tau)\delta \widetilde{u}_n(x,\tau;\beta)d\tau = 0.$$

Thus, we obtain the Euler–Lagrange equation

$$\lambda'(\tau) = 0, \tag{10.7}$$

and the natural boundary

$$1 + \lambda(\tau) = 0. \tag{10.8}$$

So, the Lagrange multiplier can be easily identified as follows:

$$\lambda = -1.$$

By substituting the identified Lagrange multiplier into Eq. (10.6), then the following variational iteration formula can be obtained:

$$\widetilde{u}_{n+1}(x,t;\beta) = \widetilde{u}_n(x,t;\beta) - \int_0^t \left\{ \begin{array}{l} D_t^\alpha \widetilde{u}_n(x,\tau;\beta) + D_x \widetilde{u}_n(x,\tau;\beta) \\ +\widetilde{u}_n(x,\tau;\beta)D_x \widetilde{u}_n(x,\tau;\beta) \\ -\widetilde{u}_n(x\tau;\beta)D_{xxx} \widetilde{u}_n(x,\tau;\beta) \\ -3D_x \widetilde{u}_n(x,\tau;\beta)D_{xx} \widetilde{u}_n(x,\tau;\beta) \\ -D_{xxt} \widetilde{u}_n(x,\tau;\beta) \end{array} \right\} d\tau. \tag{10.9}$$

Starting with an initial approximation $\widetilde{u}_0 = \widetilde{u}(x,0;\beta)$ given by Eq. (10.5) and using the variational iterational formula (10.6) gives

$$\widetilde{u}_0(x,t;\beta) = \widetilde{\delta}(\beta)e^{x/2},$$

$$\widetilde{u}_1(x,t;\beta) = \widetilde{\delta}(\beta)e^{x/2}\left(1 - \frac{1}{2}t\right),$$

$$\widetilde{u}_2(x,t;\beta) = \frac{1}{8}\widetilde{\delta}(\beta)e^{x/2}\left(8 - 9t + t^2 + 4\frac{t^{2-\alpha}}{\Gamma(3-\alpha)}\right),$$

$$\widetilde{u}_3(x,t,\beta) = \frac{1}{96}\widetilde{\delta}(\beta)e^{x/2}\left(96 - 183t + 42t^2 - 2t^3\right)$$

$$- \frac{1}{32}\widetilde{\delta}(\beta)e^{x/2}\left(\frac{16t^{3-2\alpha}}{\Gamma(4-2\alpha)} - \frac{56t^{2-\alpha}}{\Gamma(3-\alpha)} + \frac{16t^{3-\alpha}}{\Gamma(4-\alpha)}\right), \qquad (10.10)$$

and so on. In a similar manner, rest of the components $u_n(x,t;\beta)$ can easily be obtained.

Finally, the approximate solution may be written as $\widetilde{u}(x,t;\beta) = \lim\limits_{n\to\infty} \widetilde{u}_n(x,t;\beta)$.

10.3 SOLUTION BOUNDS FOR DIFFERENT INTERVAL INITIAL CONDITIONS

In this section, different interval initial conditions have been considered to find the uncertain solution bounds for interval fractional Fornberg–Whitham equation.

Case 1 Consider the interval initial condition as

$$\widetilde{u}(x,0;\beta) = \{0.4\beta + 0.8\} e^{x/2}.$$

Using Eq. (10.10), we have

$$\widetilde{u}_0(x,t;\beta) = \{0.4\beta + 0.8\} e^{x/2},$$

$$\widetilde{u}_1(x,t;\beta) = \{0.4\beta + 0.8\} e^{x/2}\left(1 - \frac{1}{2}t\right),$$

$$\widetilde{u}_2(x,t;\beta) = \frac{1}{8} \{0.4\beta + 0.8\} e^{x/2}\left(8 - 9t + t^2 + 4\frac{t^{2-\alpha}}{\Gamma(3-\alpha)}\right),$$

$$\widetilde{u}_3(x,t;\beta) = \frac{1}{96} \{0.4\beta + 0.8\} e^{x/2}(96 - 183t + 42t^2 - 2t^3)$$

$$- \frac{1}{32} \{0.4\beta + 0.8\} e^{x/2}\left(\frac{16t^{3-2\alpha}}{\Gamma(4-2\alpha)} - \frac{56t^{2-\alpha}}{\Gamma(3-\alpha)} + \frac{16t^{3-\alpha}}{\Gamma(4-\alpha)}\right).$$

$$(10.11)$$

To obtain the solution bounds, we put $\beta = 0$ and 1 for lower and upper bounds of the solution, respectively. So we get

$$\underline{u}_0(x,t;0) = \underline{u}_0(x,t) = 0.8e^{x/2},$$

$$\overline{u}_0(x,t;1) = \overline{u}_0(x,t) = 1.2e^{x/2},$$

$$\underline{u}_1(x,t;0) = \underline{u}_1(x,t) = 0.8e^{x/2}\left(1 - \frac{1}{2}t\right),$$

$$\overline{u}_1(x,t;1) = \overline{u}_1(x,t) = 1.2e^{x/2}\left(1 - \frac{1}{2}t\right),$$

$$\underline{u}_2(x,t;0) = \underline{u}_2(x,t) = \frac{1}{10}e^{x/2}\left(8 - 9t + t^2 + 4\frac{t^{2-\alpha}}{\Gamma(3-\alpha)}\right),$$

$$\overline{u}_2(x,t;1) = \overline{u}_2(x,t) = \frac{3}{20}e^{x/2}\left(8 - 9t + t^2 + 4\frac{t^{2-\alpha}}{\Gamma(3-\alpha)}\right),$$

$$\underline{u}_3(x,t;0) = \underline{u}_3(x,t) = \frac{1}{120}e^{x/2}(96 - 183t + 42t^2 - 2t^3)$$

$$- \frac{1}{40}e^{x/2}\left(\frac{16t^{3-2\alpha}}{\Gamma(4-2\alpha)} - \frac{56t^{2-\alpha}}{\Gamma(3-\alpha)} + \frac{16t^{3-\alpha}}{\Gamma(4-\alpha)}\right),$$

$$\overline{u}_3(x,t;1) = \overline{u}_3(x,t) = \frac{1}{80}e^{x/2}(96 - 183t + 42t^2 - 2t^3)$$

$$- \frac{3}{80}e^{x/2}\left(\frac{16t^{3-2\alpha}}{\Gamma(4-2\alpha)} - \frac{56t^{2-\alpha}}{\Gamma(3-\alpha)} + \frac{16t^{3-\alpha}}{\Gamma(4-\alpha)}\right),$$

and so on. In the same manner, the remaining components of the iteration (10.9) can be obtained. The approximate solution is given by $\widetilde{u}_3(x,t) = [\underline{u}_3(x,t), \overline{u}_3(x,t)]$.

One may note that the solution obtained by Abidi and Omrani (2011) is bounded by the present method (in the special case when $\alpha = 1$), and these are given in Tables 10.1 and 10.3.

TABLE 10.1   Comparison Table for Case 1 for $t = 1$ and $\alpha = 1$

x	$\underline{u}_3(x,t)$	$u_3(x,t)$ (Abidi and Omarani, 2011)	$\overline{u}_3(x,t)$
0	0.408333	0.510417	0.6125
0.5	0.52431	0.655388	0.786466
1	0.673228	0.841535	1.00984
1.5	0.864442	1.08055	1.29666
2	1.10997	1.38746	1.66495
2.5	1.42522	1.78153	2.13784
3	1.83002	2.28753	2.74503

TABLE 10.2   Comparison Table for Case 2 for $t = 1$ and $\alpha = 1$

x	$\underline{u}_3(x,t)$	$u_3(x,t)$ (Sakar and Erdogan, 2013)	$\overline{u}_3(x,t)$
0	0.510417	0.680556	0.850694
0.5	0.655388	0.873851	1.09231
1	0.841535	1.12205	1.40256
1.5	1.08055	1.44074	1.80092
2	1.38746	1.84994	2.31243
2.5	1.78153	2.37537	2.96922
3	2.28753	3.05004	3.81255

TABLE 10.3 Comparison Table for Case 1 for $x = 0$ and $\alpha = 1$

t	$\underline{u}_3(x,t)$	$u_3(x,t)$ (Abidi and Omarani, 2011)	$\overline{u}_3(x,t)$
0	0.8	1.	1.2
0.5	0.572917	0.716146	0.859375
1	0.408333	0.510417	0.6125
1.5	0.29375	0.367188	0.440625
2	0.216667	0.270833	0.325
2.5	0.164583	0.205729	0.246875
3	0.125	0.15625	0.1875

Case 2 In this case, interval initial condition has been considered as

$$\widetilde{u}(x, 0; \beta) = \left\{ \frac{2}{3}\beta + 1 \right\} e^{x/2}.$$

By applying the procedure discussed previously, we get

$$\underline{u}_0(x, t; 0) = \underline{u}_0(x, t) = e^{x/2},$$

$$\overline{u}_0(x, t; 1) = \overline{u}_0(x, t) = \frac{5}{3} e^{x/2},$$

$$\underline{u}_1(x, t; 0) = \underline{u}_1(x, t) = e^{x/2} \left(1 - \frac{1}{2}t \right),$$

$$\overline{u}_1(x, t; 1) = \overline{u}_1(x, t) = \frac{5}{3} e^{x/2} \left(1 - \frac{1}{2}t \right),$$

$$\underline{u}_2(x, t; 0) = \underline{u}_2(x, t) = \frac{1}{8} e^{x/2} \left(8 - 9t + t^2 + 4 \frac{t^{2-\alpha}}{\Gamma(3-\alpha)} \right),$$

$$\overline{u}_2(x, t; 1) = \overline{u}_2(x, t) = \frac{5}{24} e^{x/2} \left(8 - 9t + t^2 + 4 \frac{t^{2-\alpha}}{\Gamma(3-\alpha)} \right),$$

$$\underline{u}_3(x, t, 0) = \underline{u}_3(x, t) = \frac{1}{96} e^{x/2} (96 - 183t + 42t^2 - 2t^3)$$
$$- \frac{1}{32} e^{x/2} \left(\frac{16t^{3-2\alpha}}{\Gamma(4-2\alpha)} - \frac{56t^{2-\alpha}}{\Gamma(3-\alpha)} + \frac{16t^{3-\alpha}}{\Gamma(4-\alpha)} \right),$$

$$\overline{u}_3(x, t, 1) = \overline{u}_3(x, t) = \frac{5}{288} e^{x/2} (96 - 183t + 42t^2 - 2t^3)$$
$$- \frac{5}{96} e^{x/2} \left(\frac{16t^{3-2\alpha}}{\Gamma(4-2\alpha)} - \frac{56t^{2-\alpha}}{\Gamma(3-\alpha)} + \frac{16t^{3-\alpha}}{\Gamma(4-\alpha)} \right).$$

Again, rest of the components of iteration (10.9) may easily be obtained. One may note that the solution obtained by the proposed method contained the crisp solution obtained by Sakar and Erdogan (2013) and is incorporated in Tables 10.2 and 10.4.

TABLE 10.4   Comparison Table for Case 2 for $x = 0$ and $\alpha = 1$

t	$\underline{u}_3(x, t)$	$u_3(x, t)$ (Sakar and Erdogan, 2013)	$\overline{u}_3(x, t)$
0	1.	1.33333	1.66667
0.5	0.716146	0.954861	1.19358
1	0.510417	0.680556	0.850694
1.5	0.367188	0.489583	0.611979
2	0.270833	0.361111	0.451389
2.5	0.205729	0.274306	0.342882
3	0.15625	0.208333	0.260417

10.4   NUMERICAL RESULTS

Numerical results for interval fractional Fornberg–Whitham equation with different interval initial conditions are incorporated. Results of the present analysis are compared with the solutions of Abidi and Omrani (2011) and Sakar and Erdogan (2013) in special cases. Computed results are shown in terms of plots and tables.

Interval solutions for Cases 1 and 2 are depicted in Figs. 10.1 and 10.2 by varying x from 0 to 3 and keeping $t = 1$, respectively, with $\alpha = 0.4, 0.6$ and 0.8. Similarly, for both Cases 1 and 2 are shown in Figs. 10.3 and 10.4 by varying t from 0 to 3 and keeping $x = -1$ and 0, respectively, with $\alpha = 0.4, 0.6$ and 0.8. It may be worth mentioning that results for Cases 1 and 2 agree well with the solutions of Abidi and Omrani (2011) (in special case, i.e., $\alpha = 1$) and Sakar and Erdogan (2013). Comparison plots for both the cases have been shown in Figs. 10.5 and 10.6. Also, it

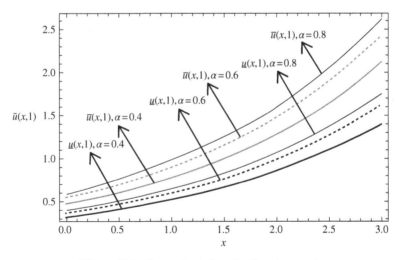

Figure 10.1   Interval solutions for Case 1 at $t = 1$

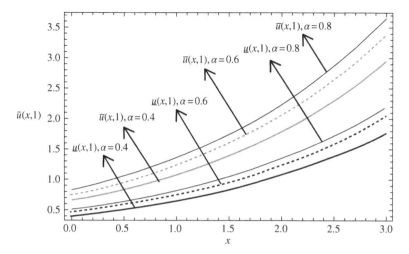

Figure 10.2   Interval solutions for Case 2 at $t = 1$

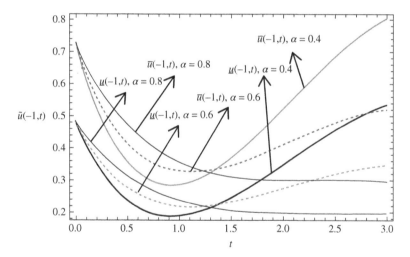

Figure 10.3   Interval solutions for Case 1 at $x = -1$

is interesting to note from Figs. 10.5 and 10.6 that the left and right bounds of the uncertain solution, that is, $\widetilde{u}(x, t)$ (with particular values of $\alpha = 1$ and $t = 1$), gradually increase with the increase of the fractional order α and time t. The approximate solutions $\widetilde{u}_3(x, t)$ for $\alpha = 0.8$ for both the cases are depicted in Figs. 10.7 and 10.8 by varying t and x from 0 to 3. Computed results are also shown in Tables 10.1 and 10.2, respectively, by varying x from 0 to 3 and keeping t constant for $\alpha = 1$. Similarly obtained results are tabulated in Tables 10.3 and 10.4 by varying t from 0 to 3 and keeping x constant for $\alpha = 1$.

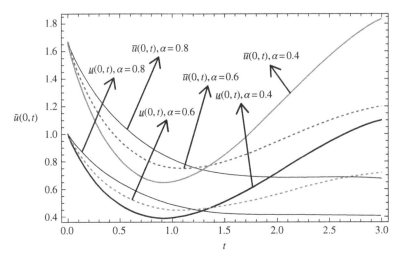

Figure 10.4   Interval solutions for Case 2 at $x = 0$ (Abidi and Omrani, 2011)

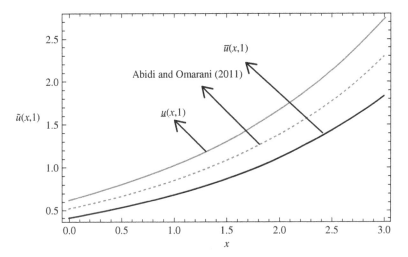

Figure 10.5   Comparison solutions for Case 1 at $t = 1$ and $\alpha = 1$

In this chapter, parametric form of interval has been applied for the solution of uncertain but bounded fractional Fornberg–Whitham equation. Uncertainties are assumed to be present in the initial conditions of the corresponding interval differential equation. VIM is used in the solution procedure, and the parametric form approach is found to be easy and straightforward.

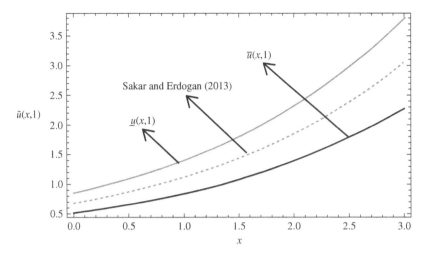

Figure 10.6 Comparison solutions for Case 2 at $t = 1$ and $\alpha = 1$

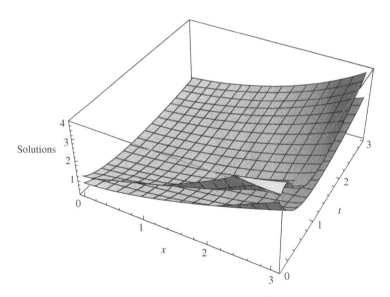

Figure 10.7 The surface shows the approximate solution $\widetilde{u}_3(x, t)$ of Case 1 for $\alpha = 0.8$

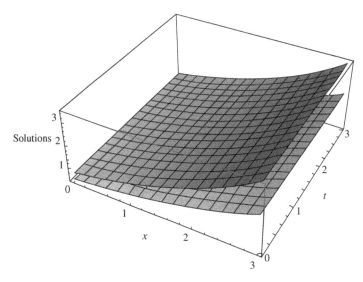

Figure 10.8   The surface shows the approximate solution $\tilde{u}_3(x, t)$ of Case 2 for $\alpha = 0.8$

BIBLIOGRAPHY

Abbaoui K, Cherruault Y. New ideas for proving convergence of decomposition methods. *Comput Math Appl* 1995;**29**:103–108.

Abdou MA, Soliman AA. Variational iteration method for solving Burger's and coupled Burger's equations. *J Comput Appl Math* 2005;**181**:245–251.

Abidi F, Omrani K. The homotopy analysis method for solving the Fornberg–Whitham equation and comparison with Adomian's decomposition method. *Comput Math Appl* 2010;**59**:2743–2750.

Abidi F, Omrani K. Numerical solution for the nonlinear Forenberg–Whitham equation by He's methods. *Int J Modern Phys B*. 2011;**25**:4721–4732.

Abulwafa EM, Abdou MA, Mahmoud AA. Variational iteration method to solve nonlinear Boltzmann equation. *Z Naturforsch* 2008;**63a**:131–139.

Chakraverty S, Tapaswini S. Non probabilistic solution of fuzzy fractional Fornberg–Whitham equation. *Comput Model Eng Sci* 2014;**103**:71–90.

Chalco-Cano Y, Rufián-Lizana A, Román-Flores H, Jiménez-Gamero MD. Calculus for interval-valued functions using generalized Hukuhara derivative and applications. *Fuzzy Sets Syst* 2013;**219**:49–67.

Fornberg B, Whitham GB. A numerical and theoretical study of certain nonlinear wave phenomena. *Phil Trans R Soc* 1978;**A289**:373–404.

He JH. Variational iteration method kind of non-linear analytical technique: Some examples. *Int J Nonlin Mech* 1999;**34**:699–708.

He JH. Variational iteration method for autonomous ordinary differential system. *Appl Math Comput* 2000;**34**:115–123.

He JH. Variational iteration method-some recent results and new interpretations. *J Comput Appl Math* 2007;**207**:3–17.

He B, Meng Q, Li S. Explicit peakon and solitary wave solutions for the modified Fornberg–Whitham equation. *Appl Math Comput* 2010;**217**:1976–1982.

Hemeda AA. Variational iteration method for solving wave equation. *Comput Math Appl* 2008;**56**:1948–1953.

Hoffmann T, Marciniak A. Solving the Poisson equation by an interval difference method of the second order. *Comput Methods Sci Tech* 2013;**19**:13–21.

Huang Y-J, Liu H-K. A new modification of the variational iteration method for van der Pol equations. *Appl Math Modell* 2013;**37**:8118–8130.

Jafari H, Tajadodi H, Baleanu D. A modified variational iteration method for solving fractional Riccati differential equation by Adomian polynomials. *Fract Calc Appl Anal.* 2013;**16**:109–122.

Lin J. Application of modified variational iteration method to the Bratu-type problems. *Int J Contemp Math Sci* 2010;**5**:153–158.

Lu J. An analytical approach to the Fornberg–Whitham type equations by using the variational iteration method. *Comput Math Appl* 2011;**61**:2010–2013.

Mahmoudi Y, Kazemian M. Some notes on homotopy analysis method for solving the Fornberg–Whitham equation. *J Basic Appl Sci Res* 2012;**2**:2985–2990.

Maidi A, Corriou JP. Open-loop optimal controller design using variational iteration method. *Appl Math Comput* 2013;**219**:8632–8645.

Merdan M, Gokdogan A, Yıldırım A, Mohyud-Din ST. Numerical simulation of fractional Fornberg–Whitham equation by differential transformation method. *Abstract Appl Anal* 2012;**2012**:1–8.

Momani S, Abuasad S, Odibat Z. Variational iteration method for solving nonlinear boundary value problems. *Appl Math Comput* 2006;**183**:1351–1358.

Moore RE. Automatic local coordinate transformations to reduce the growth of error bounds in interval computation of solution of ordinary differential equations. In: Rall LB, editor. *Error in Digital Computation II*. New York: John Wiley; 1965a. p 103–140.

Moore RE. The automatic analysis and control of error in digital computation based on the use of interval numbers. In: Rall LB, editor. *Error in Digital Computation I*. NewYork: John Wiley; 1965b. p 61–130.

Moore RE. *Interval Analysis*. Englewood Cliffs: Prentice Hall; 1966.

Moore RE. *Computational Functional Analysis*. England: Ellis Horwood Limited; 1985.

Nedialkov NS, Jackson KR. An interval Hermite-Obreschkoff method for computing rigorous bound on the solution of an initial value problem for an ordinary differential equation. *Reliab comput* 1999;**5**:289–310.

Neher M. An enclosure method for the solution of linear ODEs with polynomial coefficients. *Numer Funct Anal Optimiz* 1999;**20**:779–803.

Neher M. Geometric series bounds for the local errors of Taylor methods for linear n-th order ODEs. In: Alefeld G, Rohn J, Rump S, Yamamoto T, editors. *Symbolic Algebraic Methods and Verification Methods*. Wien: Springer; 2001. p 183–193.

Nuseir AS. New exact solutions to the modified Fornberg–Whitham equation. *Taiwanese J Math* 2012;**16**:2083–2091.

Sakar MG, Erdogan F. The homotopy analysis method for solving the time fractional Fornberg–Whitham equation and comparison with Adomian's decomposition method. *Appl Math Modell* 2013;**37**:8876–8885.

Sakar MG, Erdogan F, Yildirim A. Variational iteration method for the time-fractional Fornberg–Whitham equation. *Comput Math Appl* 2012;**63**:1382–1388.

Salama AA, Hamdy E. Interval schemes for singularly perturbed initial value problems. *Reliab Comput* 2005;**11**:41–58.

Stefanini L, Bede B. Generalized Hukuhara differentiability of interval-valued functions and interval differential equations. *Nonlin Anal Theory Methods Appl* 2009;**71**:1311–1328.

Stefanini L., Bede B. 2012. Some notes on generalized Hukuhara differentiability of interval-valued functions and interval differential equations. WP-EMS. Available at http://ideas.repec.org/f/pst233.html. Accessed 2016 Mar 26.

Wazwaz A-M. The variational iteration method for analytic treatment for linear and nonlinear ODEs. *Appl Math Comput* 2009;**212**:120–134.

Zhou J, Tian L. A type of bounded traveling wave solutions for the Fornberg–Whitham equation. *J Math Anal Appl* 2008;**346**:255–261.

Zhou J, Tian L. Periodic and solitary wave solutions to the Fornberg–Whitham equation. *Math Prob Eng* 2009;**2009**:1–10.

11

FUZZY FRACTIONAL VIBRATION EQUATION OF LARGE MEMBRANE

Vibration of large membrane has a great importance in many areas of science and engineering problems. For example, in music and acoustics, membranes constitute major components. In addition to these, membranes constitute components of microphones, speakers, and other devices. In bioengineering, many human tissues are considered as membranes. Vibration characteristics of an eardrum are important in understanding hearing. Design of hearing aid devices involves knowledge of vibration behavior of membranes.

Moreover, membranes may also be used to study two-dimensional wave mechanics and propagation. The fundamental equations of wave propagation are the same as the membrane vibration equations, namely a partial differential equation. Membranes of various shapes are being analyzed throughout the globe following different modeling aspects and computational techniques. Two spatial dimensions may be represented using the Cartesian coordinate system (usually for rectangular membranes) or using the polar coordinate system (usually for circular membranes).

Accordingly, this chapter presents the solution procedure of the governing fractional partial differential equation to handle vibration of uncertain membranes. The uncertainty has been considered as fuzzy, and double parametric form of fuzzy numbers (as discussed in Chapter 1) has been used in this problem. The fuzzy fractional partial differential equation related to the membrane is first converted to interval fuzzy fractional differential equation using single parametric form of fuzzy numbers. Then this is transformed to crisp differential equation using double parametric

Fuzzy Arbitrary Order System: Fuzzy Fractional Differential Equations and Applications, First Edition.
Snehashish Chakraverty, Smita Tapaswini, and Diptiranjan Behera.
© 2016 John Wiley & Sons, Inc. Published 2016 by John Wiley & Sons, Inc.

form of fuzzy numbers. Finally, the obtained differential equation has been solved by the methods of homotopy perturbation method (HPM) and Adomian decomposition method (ADM). For vibration of large membrane, uncertainties are assumed to be present in the initial condition and wave velocity, which are modeled through Gaussian convex normalized fuzzy sets.

11.1　DOUBLE-PARAMETRIC-BASED SOLUTION OF UNCERTAIN VIBRATION EQUATION OF LARGE MEMBRANE

Let us now consider the fuzzy fractional vibration equation of large membrane as

$$\frac{\partial^2 \tilde{v}}{\partial s^2} + \frac{1}{s}\frac{\partial \tilde{v}}{\partial s} = \frac{1}{\tilde{c}^2}\frac{\partial^\alpha \tilde{v}}{\partial t^\alpha}, \quad s \geq 0, \ t \geq 0, \ 1 < \alpha \leq 2. \tag{11.1}$$

The given equation may be written as

$$\frac{\partial^\alpha \tilde{v}}{\partial t^\alpha} = \tilde{c}^2 \left(\frac{\partial^2 \tilde{v}}{\partial s^2} + \frac{1}{s}\frac{\partial \tilde{v}}{\partial s} \right), \tag{11.2}$$

with fuzzy initial conditions

$$\tilde{v}(s,0) = (0.8, 1, 1.2)f(s), \tag{11.3}$$

$$\tilde{v}'(s,0) = \tilde{c}g(s), \tag{11.4}$$

where $\tilde{v}(s,t)$ represents the uncertain displacement and $\tilde{c}$ is the wave velocity of free vibration.

We may consider Eq. (11.2) as

$$\frac{\partial^2 \tilde{v}}{\partial t^2} = \tilde{c}^2 \frac{\partial^{2-\alpha} \tilde{v}}{\partial t^{2-\alpha}} \left(\frac{\partial^2 \tilde{v}}{\partial s^2} + \frac{1}{s}\frac{\partial \tilde{v}}{\partial s} \right), \tag{11.5}$$

where $L_{tt} \equiv \partial^2/\partial t^2$, $L_t^{2-\alpha} \equiv \partial^{2-\alpha}/\partial t^{2-\alpha}$, $L_{ss} \equiv \partial^2/\partial s^2$, and $L_s \equiv \partial/\partial s$.

As per single parametric form, we may write the given fuzzy vibration equation (Eq. (11.5)) as

$$L_{tt}\tilde{v}(s,t;r) = \left[L_{tt}\underline{v}(s,t;r), L_{tt}\overline{v}(s,t;r) \right]$$

$$= \left[\underline{c}(r), \overline{c}(r) \right]^2 L_t^{2-\alpha} \left(\begin{array}{c} \left[L_{ss}\underline{v}(s,t), L_{ss}\overline{v}(s,t) \right] \\ +\dfrac{1}{s}\left[L_s\underline{v}(s,t), L_s\overline{v}(s,t) \right] \end{array} \right), \tag{11.6}$$

subject to fuzzy initial condition

$$\left[\underline{v}(s,0;r), \overline{v}(s,0;r) \right] = [0.2r + 0.8, 1.2 - 0.2r]f(s),$$

$$\left[\underline{v}'(s,0;r), \overline{v}'(s,0;r) \right] = \left[\underline{c}(r), \overline{c}(r) \right]g(s), \quad \text{where } r \in [0,1].$$

Next, using double parametric form (as discussed in Chapter 1), Eq. (11.6) can be expressed as

$$
\begin{aligned}
&\{\beta\left(L_{tt}\overline{v}(s,t;r) - L_{tt}\underline{v}(s,t;r)\right) + L_{tt}\underline{v}(s,t;r)\} = \{\beta\left(\overline{c}(r) - \underline{c}(r)\right) + \underline{c}(r)\}^2 \\
&\{\beta\left(L_t^{2-\alpha}\overline{v}(s,t;r) - L_t^{2-\alpha}\underline{v}(s,t;r)\right) + L_t^{2-\alpha}\underline{v}(s,t;r)\} \\
&\left(\{\beta\left(L_{ss}\underline{v}(s,t;r) - L_{ss}\overline{v}(s,t;r)\right) + L_{ss}\underline{v}(s,t;r)\}\right. \\
&\left.\quad + \frac{1}{s}\{\beta\left(L_s\underline{v}(s,t;r) - L_s\overline{v}(s,t;r)\right) + L_s\underline{v}(s,t;r)\}\right),
\end{aligned}
\tag{11.7}
$$

subject to the initial conditions

$$
\{\beta\left(\underline{v}(s,0;r) - \overline{v}(s,0;r)\right) + \underline{v}(s,0;r)\} = \{\beta(0.4 - 0.4r) + (0.2r + 0.8)\}f(s),
$$

$$
\left\{\beta\left(\underline{v}'(s,0;r) - \overline{v}'(s,0;r)\right) + \underline{v}'(s,0;r)\right\} = \{\beta\left(\overline{c}(r) - \underline{c}(r)\right) + \underline{c}(r)\}g(s),
$$

where $r, \beta \in [0, 1]$.

It is now worth mentioning that Eq. (11.6) with the interval initial conditions is converted to crisp form for particular values of r and β.

Let us now denote

$$
\{\beta\left(L_{tt}\overline{v}(s,t;r) - L_{tt}\underline{v}(s,t;r)\right) + L_{tt}\underline{v}(s,t;r)\} = L_{tt}\widetilde{v}(s,t;r,\beta),
$$

$$
\{\beta\left(L_t^{2-\alpha}\overline{v}(s,t;r) - L_t^{2-\alpha}\underline{v}(s,t;r)\right) + L_t^{2-\alpha}\underline{v}(s,t;r)\} = L_t^{2-\alpha}\widetilde{v}(s,t;r,\beta),
$$

$$
\{\beta\left(L_{ss}\underline{v}(s,t;r) - L_{ss}\overline{v}(s,t;r)\right) + L_{ss}\underline{v}(s,t;r)\} = L_{ss}\widetilde{v}(s,t;r,\beta),
$$

$$
\{\beta\left(L_s\underline{v}(s,t;r) - L_s\overline{v}(s,t;r)\right) + L_s\underline{v}(s,t;r)\} = L_s\widetilde{v}(s,t;r,\beta),
$$

$$
\{\beta\left(\overline{c}(r) - \underline{c}(r)\right) + \underline{c}(r)\} = \widetilde{c}(r,\beta),
$$

$$
\{\beta\left(\underline{v}(s,0;r) - \overline{v}(s,0;r)\right) + \underline{v}(s,0;r)\} = \widetilde{v}(s,0;r,\beta) \text{ and}
$$

$$
\{\beta\left(\underline{v}'(s,0;r) - \overline{v}'(s,0;r)\right) + \underline{v}'(s,0;r)\} = \widetilde{v}'(0;r,\beta).
$$

Substituting these values in Eq. (11.7), we get

$$
L_{tt}\widetilde{v}(s,t;r,\beta) = \left(\widetilde{c}(r,\beta)\right)^2 L_t^{2-\alpha}\left(L_{ss}\widetilde{v}(s,t;r,\beta) + \frac{1}{s}L_s\widetilde{v}(s,t;r,\beta)\right),
\tag{11.8}
$$

with initial conditions

$$
\widetilde{v}(s,0;r,\beta) = \{\beta(0.4 - 0.4r) + (0.2r + 0.8)\}f(s),
$$

$$
\widetilde{v}'(s,0;r,\beta) = \{\beta\left(\overline{c}(r) - \underline{c}(r)\right) + \underline{c}(r)\}g(s).
$$

Solving the corresponding crisp differential equation, one may get the solution as $\widetilde{v}(s,t;r,\beta)$. To obtain the lower and upper bound of the solution in single parametric form, we may put $\beta = 0$ and 1, respectively. This may be represented as

$$\widetilde{v}(s, t; r, 0) = \underline{v}(s, t, r) \quad \text{and} \quad \widetilde{v}(s, t, r, 1) = \overline{v}(s, t, r).$$

Similarly, other results may be obtained by plugging in different values of r and β.

Here we have applied HPM with proposed method to obtain the solution of fuzzy vibration equations of large membranes.

11.2   SOLUTIONS OF FUZZY VIBRATION EQUATION OF LARGE MEMBRANE

- **Solution by HPM**

  According to HPM, we may now construct a simple homotopy for Eq. (11.8) with an embedding parameter $p \in [0, 1]$, as follows:

$$(1 - p) L_{tt}\widetilde{v}(s, t; r, \beta)$$

$$+ p \left[L_{tt}\widetilde{v}(s, t; r, \beta) - \left(\widetilde{c}(r, \beta)\right)^2 L_t^{2-\alpha} \left(L_{ss}\widetilde{v}(s, t; r, \beta) + \frac{1}{s} L_s\widetilde{v}(s, t; r, \beta)\right)\right] = 0$$

$$(11.9)$$

or

$$L_{tt}\widetilde{v}(s, t; r, \beta) + p \left[-\left(\widetilde{c}(r, \beta)\right)^2 L_t^{2-\alpha} \left(L_{ss}\widetilde{v}(s, t; r, \beta) + \frac{1}{s} L_s\widetilde{v}(s, t; r, \beta)\right)\right] = 0.$$

$$(11.10)$$

In the changing process from 0 to 1, for $p = 0$, Eq. (11.9) or (11.10) gives $L_{tt}\widetilde{v}(s, t; r, \beta) = 0$ and for $p = 1$, we have the original system.

In the changing process from 0 to 1, for $p = 0$, Eq. (11.9) or (11.10) gives $L_{tt}\widetilde{v}(s, t; r, \beta) = 0$ and for $p = 1$, we have the original system

$$L_{tt}\widetilde{v}(s, t; r, \beta) - \left(\widetilde{c}(r, \beta)\right)^2 L_t^{2-\alpha} \left(L_{ss}\widetilde{v}(s, t; r, \beta) + \frac{1}{s} L_s\widetilde{v}(s, t; r, \beta)\right) = 0.$$

This is called deformation in topology.

$$L_{tt}\widetilde{v}(s, t; r, \beta) \quad \text{and} \quad -\left(\widetilde{c}(r, \beta)\right)^2 L_t^{2-\alpha} \left(L_{ss}\widetilde{v}(s, t; r, \beta) + \frac{1}{s} L_s\widetilde{v}(s, t; r, \beta)\right)$$

are called homotopic. Next, we can assume the solution of Eq. (11.9) or (11.10) as a power series expansion in p as

$$\widetilde{v}(s, t; r, \beta) = \widetilde{v}_0(s, t; r, \beta) + p\widetilde{v}_1(s, t; r, \beta) + p^2\widetilde{v}_2(s, t; r, \beta)$$

$$+ p^3\widetilde{v}_3(s, t; r, \beta) + \cdots,$$

$$(11.11)$$

where, $\widetilde{v}_i(s, t; r, \beta)$ for $i = 0, 1, 2, 3, \ldots$ are functions yet to be determined. Substituting Eq. (11.11) into Eq. (11.9) or (11.10) and equating the terms with the identical powers of p, we have

$$p^0 : L_{tt}\widetilde{v}_0(s,t;r,\beta) = 0, \tag{11.12}$$

$$p^1 : L_{tt}\widetilde{v}_1(s,t;r,\beta) - \left(\widetilde{c}(r,\beta)\right)^2 L_t^{2-\alpha}\left(L_{ss}\widetilde{v}_0(s,t;r,\beta) + \frac{1}{s}L_s\widetilde{v}_0(s,t;r,\beta)\right) = 0, \tag{11.13}$$

$$p^2 : L_{tt}\widetilde{v}_2(s,t;r,\beta) - \left(\widetilde{c}(r,\beta)\right)^2 L_t^{2-\alpha}\left(L_{ss}\widetilde{v}_1(s,t;r,\beta) + \frac{1}{s}L_s\widetilde{v}_1(s,t;r,\beta)\right) = 0, \tag{11.14}$$

$$p^3 : L_{tt}\widetilde{v}_3(s,t;r,\beta) - \left(\widetilde{c}(r,\beta)\right)^2 L_t^{2-\alpha}\left(L_{ss}\widetilde{v}_2(s,t;r,\beta) + \frac{1}{s}L_s\widetilde{v}_2(s,t;r,\beta)\right) = 0, \tag{11.15}$$

and so on.

Choosing initial approximation $\widetilde{v}(s,0;r,\beta)$ and applying the operator L_{tt}^{-1} on both sides of Eqs. (11.13)–(11.15), we have

$$p^0 : \widetilde{v}_0(s,t;r,\beta) = t\widetilde{v}_0(s,0;r,\beta) + \widetilde{v}_0(s,t;r,\beta), \tag{11.16}$$

$$p^1 : \widetilde{v}_1(s,t;r,\beta) = \left(\widetilde{c}(r,\beta)\right)^2 L_t^{-\alpha}\left(L_{ss}\widetilde{v}_0(s,t;r,\beta) + \frac{1}{s}L_s\widetilde{v}_0(s,t;r,\beta)\right), \tag{11.17}$$

$$p^2 : \widetilde{v}_2(s,t;r,\beta) = \left(\widetilde{c}(r,\beta)\right)^2 L_t^{-\alpha}\left(L_{ss}\widetilde{v}_1(s,t;r,\beta) + \frac{1}{s}L_s\widetilde{v}_1(s,t;r,\beta)\right), \tag{11.18}$$

$$p^3 : \widetilde{v}_3(s,t;r,\beta) = \left(\widetilde{c}(r,\beta)\right)^2 L_t^{-\alpha}\left(L_{ss}\widetilde{v}_2(s,t;r,\beta) + \frac{1}{s}L_s\widetilde{v}_2(s,t;r,\beta)\right), \tag{11.19}$$

and so on.

One may get the approximate solution $\widetilde{v}(s,t;r,\beta) = \lim_{p\to1}\widetilde{v}(s,t;r,\beta)$, which can be expressed as

$$\widetilde{v}(s,t;r,\beta) = \widetilde{v}_0(s,t;r,\beta) + \widetilde{v}_1(s,t;r,\beta) + \widetilde{v}_2(s,t;r,\beta) + \widetilde{v}_3(s,t;r,\beta) + \cdots. \tag{11.20}$$

The series obtained by HPM converges very rapidly and only few terms are required to get the approximate solutions. The proof may be found in He (1999, 2000).

Next, we have applied ADM to solve Eq. (11.8).

- **Solution by ADM**

    Consider Eq. (11.8) and apply the operator L_{tt}^{-1} (inverse operator of L_{tt}) on both sides of Eq. (11.8).

    The equivalent expression is

$$\widetilde{v}(s,t;r,\beta) = \widetilde{v}(s,0;r,\beta) + t\widetilde{v}_t(s,0;r,\beta)$$
$$+ \left(\widetilde{c}(r,\beta)\right)^2 L_t^{-\alpha}\left(L_{ss}\widetilde{v}(s,t;r,\beta) + \frac{1}{s}L_s\widetilde{v}(s,t;r,\beta)\right), \tag{11.21}$$

where

$$L_{tt}^{-1}L_{tt}^{1}\widetilde{v}(s,t;r,\beta) = \widetilde{v}(s,t;r,\beta) - \widetilde{v}(s,0;r,\beta) - t\widetilde{v}_t(s,0;r,\beta). \quad (11.22)$$

According to ADM (Adomian 1984, 1994) we assume an infinite series solution for unknown function $\widetilde{v}(s,t;r,\beta)$ as

$$\widetilde{v}(s,t;r,\beta) = \sum_{n=0}^{\infty} \widetilde{v}_n(s,t;r,\beta), \quad (11.23)$$

where the components $\widetilde{v}_0(s,t;r,\beta), \widetilde{v}_1(s,t;r,\beta), \widetilde{v}_2(s,t;r,\beta), \ldots$ are usually determined by

$$\widetilde{v}_0(s,t;r,\beta) = \widetilde{v}(s,0;r,\beta) + t\widetilde{v}_t(s,0;r,\beta), \quad (11.24)$$

$$\widetilde{v}_1(s,t;r,\beta) = \left(\widetilde{c}(r,\beta)\right)^2 L_t^{-\alpha}\left(L_{ss}\widetilde{v}_0(s,t;r,\beta) + \frac{1}{s}L_s\widetilde{v}_0(s,t;r,\beta)\right), \quad (11.25)$$

$$\widetilde{v}_2(s,t;r,\beta) = \left(\widetilde{c}(r,\beta)\right)^2 L_t^{-\alpha}\left(L_{ss}\widetilde{v}_1(s,t;r,\beta) + \frac{1}{s}L_s\widetilde{v}_1(s,t;r,\beta)\right), \quad (11.26)$$

$$\widetilde{v}_3(s,t;r,\beta) = \left(\widetilde{c}(r,\beta)\right)^2 L_t^{-\alpha}\left(L_{ss}\widetilde{v}_2(s,t;r,\beta) + \frac{1}{s}L_s\widetilde{v}_2(s,t;r,\beta)\right), \quad (11.27)$$

and so on.

Now substituting the aforementioned terms in Eq. (11.23), one may get the approximate solution of Eq. (11.8) as follows:

$$\widetilde{v}(s,t;r,\beta) = \widetilde{v}_0(s,t;r,\beta) + \widetilde{v}_1(s,t;r,\beta) + \widetilde{v}_2(s,t;r,\beta) + \widetilde{v}_3(s,t;r,\beta) + \cdots. \quad (11.28)$$

The aforementioned series converges very rapidly (Abbaoui and Cherruault, 1994, 1995; Cherruault, 1989), and the rapid convergence means that only few terms are required to get the approximate solutions.

Now we discuss the solution bounds of the fuzzy fractional vibration equations of large membranes for particular cases. First, we consider the solution bound by HPM.

11.3   CASE STUDIES (SOLUTION BOUNDS FOR PARTICULAR CASES)

- **Using HPM**

    In this section, we have considered fuzzy initial conditions in single parametric form as $\widetilde{v}(s,0;r) = [0.2r + 0.8,\ 1.2 - 0.2r]f(s)$, $\widetilde{v}(s,0;r) = [r + 5,\ 7 - r]g(s)$ and the wave velocity as $\widetilde{c} = [r + 5, 7 - r]$. Depending upon the functions $f(s)$ and $g(s)$, we will have different cases (Yildirim et al., 2010), which are discussed in the following paragraphs for finding the uncertain solution bounds.

Case 1   Here we have taken $f(s) = s^2$ and $g(s) = s$ in the aforementioned fuzzy initial conditions. Hence, Eq. (11.8) will become

$$L_{tt}\widetilde{v}(s, t; r, \beta) = [r + 5, 7 - r]^2 \left(L_t^{2-\alpha}\left(L_{ss}\widetilde{v}(s, t; r, \beta) + \frac{1}{s}L_s\widetilde{v}(s, t; r, \beta)\right)\right).$$
(11.29)

Using double parametric form, Eq. (11.29) and the corresponding fuzzy initial conditions will become

$$L_{tt}\widetilde{v}(s, t; r, \beta)$$
$$= (\beta(2 - 2r) + (r + 5))^2 \left(L_t^{2-\alpha}\left(L_{ss}\widetilde{v}(s, t; r, \beta) + \frac{1}{s}L_s\widetilde{v}(s, t; r, \beta)\right)\right)$$

and

$$\widetilde{v}(s, 0; r, \beta) = \beta(0.4 - 0.4r) + (0.2r + 0.8)f(s),$$
$$\widetilde{v}(r, 0; \alpha, \beta) = \beta(2 - 2r) + (r + 5)g(s).$$

Let us now denote

$$\beta(0.4 - 0.4r) + (0.2r + 0.8) = \delta, \quad \beta(2 - 2r) + (r + 5) = \eta.$$

Applying HPM, we have

$$\widetilde{v}_0(s, t; r, \beta) = \eta st + \delta s^2,$$
(11.30)

$$\widetilde{v}_1(s, t; r, \beta) = 4\delta\eta^2\frac{t^\alpha}{\Gamma(\alpha + 1)} + \frac{\eta^3}{s}\frac{t^{\alpha+1}}{\Gamma(\alpha + 2)},$$
(11.31)

$$\widetilde{v}_2(s, t; r, \beta) = \frac{\eta^5}{s^3}\frac{t^{2\alpha+1}}{\Gamma(2\alpha + 2)},$$
(11.32)

$$\widetilde{v}_3(s, t; r, \beta) = \frac{9\eta^7}{s^5}\frac{t^{3\alpha+1}}{\Gamma(3\alpha + 2)},$$
(11.33)

and so on.

Therefore, the solution can be written as

$$\widetilde{v}(s, t; r, \beta) = s^2 \left(\begin{array}{l} \delta + t\dfrac{\eta}{s} + 4\delta\dfrac{t^\alpha}{\Gamma(\alpha + 1)}\left(\dfrac{\eta}{s}\right)^2 + \dfrac{t^{\alpha+1}}{\Gamma(\alpha + 2)}\left(\dfrac{\eta}{s}\right)^3 + \\[3mm] \dfrac{t^{2\alpha+1}}{\Gamma(2\alpha + 2)}\left(\dfrac{\eta}{s}\right)^5 + 9\dfrac{t^{3\alpha+1}}{\Gamma(3\alpha + 2)}\left(\dfrac{\eta}{s}\right)^7 + \cdots \end{array}\right)$$
(11.34)

or

$$\tilde{v}(s,t;r,\beta) = s^2 \begin{pmatrix} (\beta\,(0.4-0.4r)+(0.2r+0.8)) + \left(\dfrac{(\beta\,(2-2r)+(r+5))}{s}\right)t \\[2mm] +4\,(\beta\,(0.4-0.4r)+(0.2r+0.8))\left(\dfrac{(\beta\,(2-2r)+(r+5))}{s}\right)^2 \\[2mm] \times\dfrac{t^\alpha}{\Gamma(\alpha+1)} + \left(\dfrac{(\beta\,(2-2r)+(r+5))}{s}\right)^3\dfrac{t^{\alpha+1}}{\Gamma(\alpha+2)} \\[2mm] +\left(\dfrac{(\beta\,(2-2r)+(r+5))}{s}\right)^5\dfrac{t^{2\alpha+1}}{\Gamma(2\alpha+2)} \\[2mm] +9\left(\dfrac{(\beta\,(2-2r)+(r+5))}{s}\right)^7\dfrac{t^{3\alpha+1}}{\Gamma(3\alpha+2)} + \cdots \end{pmatrix}.$$

$$(11.35)$$

To obtain the solution bounds in single parametric form, we may substitute $\beta = 0$ and 1 in Eq. (11.35) for lower and upper bounds of the solution, respectively. So we get

$$\underline{v}(s,t;r,0) = s^2 \begin{pmatrix} (0.2r+0.8) + \left(\dfrac{(r+5)}{s}\right)t \\[2mm] +4\,(0.2r+0.8)\left(\dfrac{(r+5)}{s}\right)^2\dfrac{t^\alpha}{\Gamma(\alpha+1)} \\[2mm] +\left(\dfrac{(r+5)}{s}\right)^3\dfrac{t^{\alpha+1}}{\Gamma(\alpha+2)} + \left(\dfrac{(r+5)}{s}\right)^5\dfrac{t^{2\alpha+1}}{\Gamma(2\alpha+2)} \\[2mm] +9\left(\dfrac{(r+5)}{s}\right)^7\dfrac{t^{3\alpha+1}}{\Gamma(3\alpha+2)} + \cdots \end{pmatrix} \qquad (11.36)$$

and

$$\tilde{v}(s,t;r,1) = s^2 \begin{pmatrix} (1.2-0.2r) + \left(\dfrac{(7-r)}{s}\right)t \\[2mm] +4\,(1.2-0.2r)\left(\dfrac{(7-r)}{s}\right)^2\dfrac{t^\alpha}{\Gamma(\alpha+1)} \\[2mm] +\left(\dfrac{(7-r)}{s}\right)^3\dfrac{t^{\alpha+1}}{\Gamma(\alpha+2)} + \left(\dfrac{(7-r)}{s}\right)^5\dfrac{t^{2\alpha+1}}{\Gamma(2\alpha+2)} \\[2mm] +9\left(\dfrac{(7-r)}{s}\right)^7\dfrac{t^{3\alpha+1}}{\Gamma(3\alpha+2)} + \cdots \end{pmatrix}. \qquad (11.37)$$

One may note that in the special case when $r = 1$ and wave velocity $c = 6$, the crisp results obtained by the proposed method are exactly same as that of the solution obtained by Yildirim et al. (2010) for $\alpha = 2$. The aforementioned series will be convergent for the values of $|t/s| \leq 1$, that is, for large membrane and small range of time.

Case 2   We consider $f(s) = s$ and $g(s) = 1$.

Again, by applying the procedure discussed previously, we get the solution

$$\tilde{v}_0(s,t;r,\beta) = \eta t + \delta s^2, \tag{11.38}$$

$$\tilde{v}_1(s,t;r,\beta) = \frac{\eta^2 \delta}{s} \frac{t^\alpha}{\Gamma(\alpha+1)}, \tag{11.39}$$

$$\tilde{v}_2(s,t;r,\beta) = \frac{\eta^4 \delta}{s^3} \frac{t^{2\alpha}}{\Gamma(2\alpha+1)}, \tag{11.40}$$

$$\tilde{v}_3(s,t;r,\beta) = \frac{9\eta^6 \delta}{s^5} \frac{t^{3\alpha}}{\Gamma(3\alpha+1)}, \tag{11.41}$$

and so on.

The solution may be obtained as

$$\tilde{v}(s,t;r,\beta) = \eta t + \delta s + \frac{\eta^2 \delta}{s} \frac{t^\alpha}{\Gamma(\alpha+1)} + \frac{\eta^4 \delta}{s^3} \frac{t^{2\alpha}}{\Gamma(2\alpha+1)}$$

$$+ \frac{9\eta^6 \delta}{s^5} \frac{t^{3\alpha}}{\Gamma(3\alpha+1)} + \cdots \tag{11.42}$$

or

$$\tilde{v}(s,t;r,\beta) = (\beta(2-2r) + (r+5))t + (\beta(0.4-0.4r) + (0.2r+0.8))s$$

$$+ \frac{(\beta(2-2r) + (r+5))^2 (\beta(0.4-0.4r) + (0.2r+0.8))}{s} \frac{t^\alpha}{\Gamma(\alpha+1)}$$

$$+ \frac{(\beta(2-2r) + (r+5))^4 (\beta(0.4-0.4r) + (0.2r+0.8))}{s^3} \frac{t^{2\alpha}}{\Gamma(2\alpha+1)}$$

$$+ \frac{9(\beta(2-2r) + (r+5))^6 (\beta(0.4-0.4r) + (0.2r+0.8))}{s^5}$$

$$\times \frac{t^{3\alpha}}{\Gamma(3\alpha+1)} + \cdots \tag{11.43}$$

Putting $\beta = 0$ and 1 in $\tilde{v}(s,t;r,\beta)$, we get the lower and upper bounds of the fuzzy solutions, respectively, as

$$\underline{v}(s,t;r,0) = (r+5)t + (0.2r+0.8)s + \frac{(r+5)^2 (0.2r+0.8)}{s} \frac{t^\alpha}{\Gamma(\alpha+1)}$$

$$+ \frac{(r+5)^4 (0.2r+0.8)}{s^3} \frac{t^{2\alpha}}{\Gamma(2\alpha+1)}$$

$$+ \frac{9(r+5)^6 (0.2r+0.8)}{s^5} \frac{t^{3\alpha}}{\Gamma(3\alpha+1)} + \cdots \tag{11.44}$$

and

$$\widetilde{v}(s,t;r,1) = (7-r)t + (1.2 - 0.2r)s + \frac{(7-r)^2 (1.2-0.2r)}{s} \frac{t^\alpha}{\Gamma(\alpha+1)}$$

$$+ \frac{(7-r)^4 (1.2-0.2r)}{s^3} \frac{t^{2\alpha}}{\Gamma(2\alpha+1)}$$

$$+ \frac{9(7-r)^6 (1.2-0.2r)}{s^5} \frac{t^{3\alpha}}{\Gamma(3\alpha+1)} + \cdots \qquad (11.45)$$

Solution obtained by proposed method for $r = 1$ and the wave velocity $c = 6$ is again found to be exactly the same as that of (crisp result) Yildirim et al. (2010) for $\alpha = 2$.

Case 3 $f(s) = \sqrt{s}$ and $g(s) = 1/\sqrt{s}$.

By following the proposed method with HPM, we get the solution in double parametric form as

$$\widetilde{v}(s,t;r,\beta) = \begin{pmatrix} \dfrac{\eta}{\sqrt{s}} t + \delta\sqrt{s} + \dfrac{\eta^3}{4s^{5/2}} \dfrac{t^{\alpha+1}}{\Gamma(\alpha+2)} + \dfrac{\eta^2\delta}{4s^{3/2}} \dfrac{t^\alpha}{\Gamma(\alpha+1)} \\[2ex] + \dfrac{25}{16} \dfrac{\eta^5}{s^{9/2}} \dfrac{t^{2\alpha+1}}{\Gamma(2\alpha+2)} + \dfrac{9}{16} \dfrac{\eta^4\delta}{s^{7/2}} \dfrac{t^{2\alpha}}{\Gamma(2\alpha+1)} \\[2ex] + \dfrac{2025}{64} \dfrac{\eta^7}{s^{13/2}} \dfrac{t^{3\alpha+1}}{\Gamma(3\alpha+2)} + \dfrac{441}{64} \dfrac{\eta^6\delta}{s^{11/2}} \dfrac{t^{3\alpha}}{\Gamma(3\alpha+1)} + \cdots \end{pmatrix}.$$
$$(11.46)$$

The lower and upper bounds of the fuzzy solutions may again be written as

$$\underline{v}(s,t;r,0) = \begin{pmatrix} \dfrac{(r+5)}{\sqrt{s}} t + (0.2r+0.8)\sqrt{s} + \dfrac{(r+5)^3}{4s^{5/2}} \dfrac{t^{\alpha+1}}{\Gamma(\alpha+2)} \\[2ex] + \dfrac{(r+5)^2 (0.2r+0.8)}{4s^{3/2}} \dfrac{t^\alpha}{\Gamma(\alpha+1)} + \dfrac{25}{16} \dfrac{(r+5)^5}{s^{9/2}} \dfrac{t^{2\alpha+1}}{\Gamma(2\alpha+2)} \\[2ex] + \dfrac{9}{16} \dfrac{(r+5)^4 (0.2r+0.8)}{s^{7/2}} \dfrac{t^{2\alpha}}{\Gamma(2\alpha+1)} + \dfrac{2025}{64} \dfrac{(r+5)^7}{s^{13/2}} \dfrac{t^{3\alpha+1}}{\Gamma(3\alpha+2)} \\[2ex] + \dfrac{441}{64} \dfrac{(r+5)^6 (0.2r+0.8)}{s^{11/2}} \dfrac{t^{3\alpha}}{\Gamma(3\alpha+1)} + \cdots \end{pmatrix}$$
$$(11.47)$$

and

$$\overline{v}(s,t;r,1) = \begin{pmatrix} \dfrac{(7-r)}{\sqrt{s}} t + (1.2-0.2r)\sqrt{s} + \dfrac{(7-r)^3}{4s^{5/2}} \dfrac{t^{\alpha+1}}{\Gamma(\alpha+2)} \\[2ex] + \dfrac{(7-r)^2 (1.2-0.2r)}{4s^{3/2}} \dfrac{t^\alpha}{\Gamma(\alpha+1)} + \dfrac{25}{16} \dfrac{(7-r)^5}{s^{9/2}} \dfrac{t^{2\alpha+1}}{\Gamma(2\alpha+2)} \\[2ex] + \dfrac{9}{16} \dfrac{(7-r)^4 (1.2-0.2r)}{s^{7/2}} \dfrac{t^{2\alpha}}{\Gamma(2\alpha+1)} + \dfrac{2025}{64} \dfrac{(7-r)^7}{s^{13/2}} \dfrac{t^{3\alpha+1}}{\Gamma(3\alpha+2)} \\[2ex] + \dfrac{441}{64} \dfrac{(7-r)^6 (1.2-0.2r)}{s^{11/2}} \dfrac{t^{3\alpha}}{\Gamma(3\alpha+1)} + \cdots \end{pmatrix}.$$
$$(11.48)$$

Case 4 $f(s) = s^2$ and $g(s) = 1$.

In this case, we have

$$\tilde{v}_0(s, t; r, \beta) = \eta t + \delta s^2, \tag{11.49}$$

$$\tilde{v}_1(s, t; r, \beta) = 4\delta\eta^2 \frac{t^\alpha}{\Gamma(\alpha + 1)}, \tag{11.50}$$

$$\tilde{v}_2(s, t; r, \beta) = 0, \tag{11.51}$$

$$\tilde{v}_n(s, t; r, \beta) = 0, \quad n \geq 2. \tag{11.52}$$

Therefore, the solution in double parametric form is as follows:

$$\tilde{v}(s, t; r, \beta) = \eta t + \delta s^2 + 4\delta\eta^2 \frac{t^\alpha}{\Gamma(\alpha + 1)}, \tag{11.53}$$

where the lower and upper bounds of the fuzzy solutions are obtained as

$$\underline{v}(s, t; r, 0) = (r + 5) t + (0.2r + 0.8) s^2 + 4(0.2r + 0.8)(r + 5)^2 \frac{t^\alpha}{\Gamma(\alpha + 1)}, \tag{11.54}$$

and

$$\overline{v}(s, t; r, 1) = (7 - r) t + (1.2 - 0.2r) s^2 + 4(1.2 - 0.2r)(7 - r)^2 \frac{t^\alpha}{\Gamma(\alpha + 1)}. \tag{11.55}$$

Again one may see that the solution obtained by proposed method for $r = 1$, $\alpha = 2$ and the wave velocity $c = 6$ exactly agrees with the solution of Yildirim et al. (2010).

Case 5 In this case, we consider $f(s) = s^2$ and $g(s) = s^2$.

We have the solutions in this case as

$$\tilde{v}_0(s, t; r, \beta) = \eta s^2 t + \delta s^2, \tag{11.56}$$

$$\tilde{v}_1(s, t; r, \beta) = 2\eta^3 \frac{t^{\alpha+1}}{\Gamma(\alpha + 2)} + 4\delta\eta^2 \frac{t^\alpha}{\Gamma(\alpha + 1)}, \tag{11.57}$$

$$\tilde{v}_2(s, t; r, \beta) = 0, \tag{11.58}$$

$$\tilde{v}_n(s, t; r, \beta) = 0, \quad \text{for } n \geq 2, \tag{11.59}$$

and finally one may write

$$\tilde{v}(s, t; r, \beta) = \eta s^2 t + \delta s^2 + 2\eta^3 \frac{t^{\alpha+1}}{\Gamma(\alpha + 2)} + 4\delta\eta^2 \frac{t^\alpha}{\Gamma(\alpha + 1)}, \tag{11.60}$$

where lower and upper bound of the solutions, respectively, are

$$\underline{v}(s, t; r, 0) = (r + 5) s^2 t + (0.2r + 0.8) s^2 + 2(r + 5)^3 \frac{t^{\alpha+1}}{\Gamma(\alpha + 2)}$$
$$+ 4(0.2r + 0.8)(r + 5)^2 \frac{t^\alpha}{\Gamma(\alpha + 1)}, \tag{11.61}$$

and

$$\bar{v}(s,t;r,1) = (7-r)s^2t + (1.2-0.2r)s^2 + 2(7-r)^3\frac{t^{\alpha+1}}{\Gamma(\alpha+2)}$$

$$+ 4(1.2-0.2r)(7-r)^2\frac{t^\alpha}{\Gamma(\alpha+1)}. \quad (11.62)$$

- Using ADM

  In this section, we have considered fuzzy initial conditions in single parametric form as

$$\widetilde{v}(s,0;r) = \left[1 - 0.2\sqrt{-2\log_e r}, 1 + 0.2\sqrt{-2\log_e r}\right]f(s),$$

$$\widetilde{v}(s,0;r) = \left[6 - 0.2\sqrt{-2\log_e r}, 6 + 0.2\sqrt{-2\log_e r}\right]g(s),$$

and the wave velocity as $\widetilde{c} = \left[6 - 0.2\sqrt{-2\log_e r}, 6 + 0.2\sqrt{-2\log_e r}\right]$. Depending upon the function $f(s)$ and $g(s)$, we will have different cases Yildirim et al. (2010), which are discussed in the following paragraphs for finding the uncertain solution bounds.

Case 6 Here we have taken $f(s) = s^2$ and $g(s) = s$ in the aforementioned fuzzy initial conditions. Hence, Eq. (7.5) will become

$$L_{tt}\widetilde{v}(s,t) = \left[6 - 0.2\sqrt{-2\log_e r}, 6 + 0.2\sqrt{-2\log_e r}\right]^2$$

$$\times L_t^{2-\alpha}\left(L_{ss}\widetilde{v}(s,t) + \frac{1}{s}L_s\widetilde{v}(s,t)\right). \quad (11.63)$$

Using double parametric form, Eq. (11.63) and the corresponding fuzzy initial conditions will become

$$L_{tt}\widetilde{v}(s,t;r,\beta) = \left(\beta\left(0.4\sqrt{-2\log_e r}\right) + \left(6 - 0.2\sqrt{-2\log_e r}\right)\right)^2$$

$$\times L_t^{2-\alpha}\left(L_{ss}\widetilde{v}(s,t;r,\beta) + \frac{1}{s}L_s\widetilde{v}(s,t;r,\beta)\right)$$

and

$$\widetilde{v}(s,0;r,\beta) = \beta\left(0.4\sqrt{-2\log_e r}\right) + \left(1 - 0.2\sqrt{-2\log_e r}\right)f(s),$$

$$\widetilde{v}(s,0;r,\beta) = \beta\left(0.4\sqrt{-2\log_e r}\right) + \left(6 - 0.2\sqrt{-2\log_e r}\right)g(s).$$

Let us now denote

$$\beta\left(0.4\sqrt{-2\log_e r}\right) + \left(1 - 0.2\sqrt{-2\log_e r}\right) = \delta,$$

$$\beta\left(0.4\sqrt{-2\log_e r}\right) + \left(6 - 0.2\sqrt{-2\log_e r}\right) = \eta.$$

Applying ADM, we have

$$\tilde{v}_0(s, t; r, \beta) = \eta s t + \delta s^2, \tag{11.64}$$

$$\tilde{v}_1(s, t; r, \beta) = 4\delta\eta^2 \frac{t^\alpha}{\Gamma(\alpha+1)} + \frac{\eta^3}{s}\frac{t^{\alpha+1}}{\Gamma(\alpha+2)}, \tag{11.65}$$

$$\tilde{v}_2(s, t; r, \beta) = \frac{\eta^5}{s^3}\frac{t^{2\alpha+1}}{\Gamma(2\alpha+2)}, \tag{11.66}$$

$$\tilde{v}_3(s, t; r, \beta) = \frac{9\eta^7}{s^5}\frac{t^{3\alpha+1}}{\Gamma(3\alpha+2)}, \tag{11.67}$$

and so on.

In a similar manner, higher-order approximation may be obtained as discussed earlier.

Therefore, the solution can be written as

$$\tilde{v}(s, t; r, \beta) = s^2 \left(\begin{array}{l} \delta + t\dfrac{\eta}{s} + 4\delta\dfrac{t^\alpha}{\Gamma(\alpha+1)}\left(\dfrac{\eta}{s}\right)^2 + \dfrac{t^{\alpha+1}}{\Gamma(\alpha+2)}\left(\dfrac{\eta}{s}\right)^3 \\ + \dfrac{t^{2\alpha+1}}{\Gamma(2\alpha+2)}\left(\dfrac{\eta}{s}\right)^5 + 9\dfrac{t^{3\alpha+1}}{\Gamma(3\alpha+2)}\left(\dfrac{\eta}{s}\right)^7 + \cdots \end{array} \right). \tag{11.68}$$

To obtain the solution bounds in single parametric form, we may put $\beta = 0$ and 1 in Eq. (11.68) for lower and upper bounds of the solution, respectively. So we get

$$\underline{\tilde{v}}(s, t; r, 0) = s^2 \left| \begin{array}{l} (1 - 0.2\sqrt{-2\log_e r}) + t\dfrac{(6 - 0.2\sqrt{-2\log_e r})}{s} \\[2mm] + 4(1 - 0.2\sqrt{-2\log_e r})\dfrac{t^\alpha}{\Gamma(\alpha+1)}\left(\dfrac{(6 - 0.2\sqrt{-2\log_e r})}{s}\right)^2 \\[2mm] + \dfrac{t^{\alpha+1}}{\Gamma(\alpha+2)}\left(\dfrac{(6 - 0.2\sqrt{-2\log_e r})}{s}\right)^3 \\[2mm] + \dfrac{t^{2\alpha+1}}{\Gamma(2\alpha+2)}\left(\dfrac{(6 - 0.2\sqrt{-2\log_e r})}{s}\right)^5 \\[2mm] + 9\dfrac{t^{3\alpha+1}}{\Gamma(3\alpha+2)}\left(\dfrac{(6 - 0.2\sqrt{-2\log_e r})}{s}\right)^7 + \cdots \end{array} \right. \tag{11.69}$$

and

$$\widetilde{v}(s,t;r,\beta) = s^2 \begin{pmatrix} (1 + 0.2\sqrt{-2\log_e r}) + t\dfrac{(6 + 0.2\sqrt{-2\log_e r})}{s} \\[2mm] + 4(1 + 0.2\sqrt{-2\log_e r})\dfrac{t^\alpha}{\Gamma(\alpha + 1)}\left(\dfrac{(6 + 0.2\sqrt{-2\log_e r})}{s}\right)^2 \\[2mm] + \dfrac{t^{\alpha+1}}{\Gamma(\alpha + 2)}\left(\dfrac{(6 + 0.2\sqrt{-2\log_e r})}{s}\right)^3 \\[2mm] + \dfrac{t^{2\alpha+1}}{\Gamma(2\alpha + 2)}\left(\dfrac{(6 + 0.2\sqrt{-2\log_e r})}{s}\right)^5 \\[2mm] + 9\dfrac{t^{3\alpha+1}}{\Gamma(3\alpha + 2)}\left(\dfrac{(6 + 0.2\sqrt{-2\log_e r})}{s}\right)^7 + \cdots \end{pmatrix}.$$

$$(11.70)$$

It is interesting to note that for $r = 1$ and wave velocity $c = 6$, the crisp results obtained by the proposed method are exactly the same as that of the solution obtained by Yildirim et al. (2010) for $\alpha = 2$. The aforementioned series will be convergent for the values of $|t/s| \leq 1$, that is, for large membrane and small range of time.

Case 7 $f(s) = s$ and $g(s) = 1$.

Again, by applying the procedure discussed previously, we get the solution

$$\widetilde{v}_0(s,t;r,\beta) = \eta t + \delta s, \tag{11.71}$$

$$\widetilde{v}_1(s,t;r,\beta) = \frac{\eta^2\delta}{s}\frac{t^\alpha}{\Gamma(\alpha + 1)}, \tag{11.72}$$

$$\widetilde{v}_2(s,t;r,\beta) = \frac{\eta^4\delta}{s^3}\frac{t^{2\alpha}}{\Gamma(2\alpha + 1)}, \tag{11.73}$$

$$\widetilde{v}_3(s,t;r,\beta) = \frac{9\eta^6\delta}{s^5}\frac{t^{3\alpha}}{\Gamma(3\alpha + 1)}, \tag{11.74}$$

and so on.

The solution in general form may be obtained as

$$\widetilde{v}(s,t;r,\beta) = \eta t + \delta s + \frac{\eta^2\delta}{s}\frac{t^\alpha}{\Gamma(\alpha + 1)} + \frac{\eta^4\delta}{s^3}\frac{t^{2\alpha}}{\Gamma(2\alpha + 1)}$$

$$+ \frac{9\eta^6\delta}{s^5}\frac{t^{3\alpha}}{\Gamma(3\alpha + 1)} + \cdots \tag{11.75}$$

Putting $\beta = 0$ and 1 in $\widetilde{v}(s, t; r, \beta)$ we get the lower and upper bounds of the fuzzy solutions, respectively, as

$$
\begin{aligned}
\widetilde{v}(s, t; r, 0) = {} & (6 - 0.2\sqrt{-2\log_e r})t + (1 - 0.2\sqrt{-2\log_e r})s \\
& + \frac{(6 - 0.2\sqrt{-2\log_e r})^2(1 - 0.2\sqrt{-2\log_e r})}{s} \frac{t^\alpha}{\Gamma(\alpha + 1)} \\
& + \frac{(6 - 0.2\sqrt{-2\log_e r})^4(1 - 0.2\sqrt{-2\log_e r})}{s^3} \frac{t^{2\alpha}}{\Gamma(2\alpha + 1)} \\
& + \frac{9(6 - 0.2\sqrt{-2\log_e r})^6(1 - 0.2\sqrt{-2\log_e r})}{s^5} \frac{t^{3\alpha}}{\Gamma(3\alpha + 1)} + \cdots
\end{aligned}
$$

$$(11.76)$$

and

$$
\begin{aligned}
\widetilde{v}(s, t; r, 1) = {} & (6 + 0.2\sqrt{-2\log_e r})t + (1 + 0.2\sqrt{-2\log_e r})s \\
& + \frac{(6 + 0.2\sqrt{-2\log_e r})^2(1 + 0.2\sqrt{-2\log_e r})}{s} \frac{t^\alpha}{\Gamma(\alpha + 1)} \\
& + \frac{(6 + 0.2\sqrt{-2\log_e r})^4(1 + 0.2\sqrt{-2\log_e r})}{s^3} \frac{t^{2\alpha}}{\Gamma(2\alpha + 1)} \\
& + \frac{9(6 + 0.2\sqrt{-2\log_e r})^6(1 + 0.2\sqrt{-2\log_e r})}{s^5} \frac{t^{3\alpha}}{\Gamma(3\alpha + 1)} + \cdots
\end{aligned}
$$

$$(11.77)$$

We observe that for $r = 1$ and the wave velocity $c = 6$, the solution is again found to be exactly same as that of (crisp result) Yildirim et al. (2010) for $\alpha = 2$.

Case 8 $f(s) = \sqrt{s}$ and $g(s) = 1/\sqrt{s}$.

By following the proposed method with ADM, we get the solution in double parametric form as

$$
\widetilde{v}(s, t; r, \beta) = \begin{pmatrix}
\dfrac{\eta}{\sqrt{s}}t + \delta\sqrt{s} + \dfrac{\eta^3}{4s^{5/2}} \dfrac{t^{\alpha+1}}{\Gamma(\alpha+2)} + \dfrac{\eta^2\delta}{4s^{3/2}} \dfrac{t^\alpha}{\Gamma(\alpha+1)} \\[2ex]
+ \dfrac{25}{16} \dfrac{\eta^5}{s^{9/2}} \dfrac{t^{2\alpha+1}}{\Gamma(2\alpha+2)} + \dfrac{9}{16} \dfrac{\eta^4\delta}{s^{7/2}} \dfrac{t^{2\alpha}}{\Gamma(2\alpha+1)} \\[2ex]
+ \dfrac{2025}{64} \dfrac{\eta^7}{s^{13/2}} \dfrac{t^{3\alpha+1}}{\Gamma(3\alpha+2)} + \dfrac{441}{64} \dfrac{\eta^6\delta}{s^{11/2}} \dfrac{t^{3\alpha}}{\Gamma(3\alpha+1)} + \cdots
\end{pmatrix}
$$

$$(11.78)$$

The lower and upper bounds of the fuzzy solutions may again be written as

$$\underline{v}(s,t;r,0) = \begin{pmatrix} \dfrac{(6-0.2\sqrt{-2\log_e r})}{\sqrt{s}}t + (1-0.2\sqrt{-2\log_e r})\sqrt{s} \\[2ex] +\dfrac{(6-0.2\sqrt{-2\log_e r})^3}{4s^{5/2}}\dfrac{t^{\alpha+1}}{\Gamma(\alpha+2)} \\[2ex] +\dfrac{(6-0.2\sqrt{-2\log_e r})^2(1-0.2\sqrt{-2\log_e r})}{4s^{3/2}}\dfrac{t^{\alpha}}{\Gamma(\alpha+1)} \\[2ex] +\dfrac{25}{16}\dfrac{(6-0.2\sqrt{-2\log_e r})^5}{s^{9/2}}\dfrac{t^{2\alpha+1}}{\Gamma(2\alpha+2)} \\[2ex] +\dfrac{9}{16}\dfrac{(6-0.2\sqrt{-2\log_e r})^4(1-0.2\sqrt{-2\log_e r})}{s^{7/2}}\dfrac{t^{2\alpha}}{\Gamma(2\alpha+1)} \\[2ex] +\dfrac{2025}{64}\dfrac{(6-0.2\sqrt{-2\log_e r})^7}{s^{13/2}}\dfrac{t^{3\alpha+1}}{\Gamma(3\alpha+2)} \\[2ex] +\dfrac{441}{64}\dfrac{(6-0.2\sqrt{-2\log_e r})^6(1-0.2\sqrt{-2\log_e r})}{s^{11/2}}\dfrac{t^{3\alpha}}{\Gamma(3\alpha+1)}+\cdots \end{pmatrix}$$

(11.79)

and

$$\overline{v}(s,t;r,1) = \begin{pmatrix} \dfrac{(6+0.2\sqrt{-2\log_e r})}{\sqrt{s}}t + (1+0.2\sqrt{-2\log_e r})\sqrt{s} \\[2ex] +\dfrac{(6+0.2\sqrt{-2\log_e r})^3}{4s^{5/2}}\dfrac{t^{\alpha+1}}{\Gamma(\alpha+2)} \\[2ex] +\dfrac{(6+0.2\sqrt{-2\log_e r})^2(1+0.2\sqrt{-2\log_e r})}{4s^{3/2}}\dfrac{t^{\alpha}}{\Gamma(\alpha+1)} \\[2ex] +\dfrac{25}{16}\dfrac{(6+0.2\sqrt{-2\log_e r})^5}{s^{9/2}}\dfrac{t^{2\alpha+1}}{\Gamma(2\alpha+2)} \\[2ex] +\dfrac{9}{16}\dfrac{(6+0.2\sqrt{-2\log_e r})^4(1+0.2\sqrt{-2\log_e r})}{s^{7/2}}\dfrac{t^{2\alpha}}{\Gamma(2\alpha+1)} \\[2ex] +\dfrac{2025}{64}\dfrac{(6+0.2\sqrt{-2\log_e r})^7}{s^{13/2}}\dfrac{t^{3\alpha+1}}{\Gamma(3\alpha+2)} \\[2ex] +\dfrac{441}{64}\dfrac{(6+0.2\sqrt{-2\log_e r})^6(1+0.2\sqrt{-2\log_e r})}{s^{11/2}}\dfrac{t^{3\alpha}}{\Gamma(3\alpha+1)}+\cdots \end{pmatrix}.$$

(11.80)

Case 9 $f(s) = s^2$ and $g(s) = 1$.

In this case, we have

$$\widetilde{v}_0(s,t;r,\beta) = \eta t + \delta s^2,$$

(11.81)

$$\tilde{v}_1(s,t;r,\beta) = 4\delta\eta^2 \frac{t^\alpha}{\Gamma(\alpha+1)}, \tag{11.82}$$

$$\tilde{v}_2(s,t;r,\beta) = 0, \tag{11.83}$$

$$\tilde{v}_n(s,t;r,\beta) = 0, \quad n \geq 2. \tag{11.84}$$

Therefore, the solution in double parametric form is as follows:

$$\tilde{v}(s,t;r,\beta) = \eta t + \delta s^2 + 4\delta\eta^2 \frac{t^\alpha}{\Gamma(\alpha+1)}. \tag{11.85}$$

The lower and upper bounds of the fuzzy solutions are obtained as

$$\underline{v}(s,t;r,0) = \left(6 - 0.2\sqrt{-2\log_e r}\right)t + \left(1 - 0.2\sqrt{-2\log_e r}\right)s^2$$
$$+ 4\left(1 - 0.2\sqrt{-2\log_e r}\right)\left(6 - 0.2\sqrt{-2\log_e r}\right)^2 \frac{t^\alpha}{\Gamma(\alpha+1)} \tag{11.86}$$

and

$$\overline{v}(s,t;r,1) = \left(6 + 0.2\sqrt{-2\log_e r}\right)t + \left(1 + 0.2\sqrt{-2\log_e r}\right)s^2$$
$$+ 4\left(1 + 0.2\sqrt{-2\log_e r}\right)\left(6 + 0.2\sqrt{-2\log_e r}\right)^2 \frac{t^\alpha}{\Gamma(\alpha+1)}. \tag{11.87}$$

Again one may see that the solution obtained by proposed method for $r = 1$ and the wave velocity $c = 6$ exactly agrees with the solution of Yildirim et al. (2010) for $\alpha = 2$.

Case 10 Finally, we have considered $f(s) = s^2$ and $g(s) = s^2$.
We have the solutions in this case as

$$\tilde{v}_0(s,t;r,\beta) = \eta s^2 t + \delta s^2, \tag{11.88}$$

$$\tilde{v}_1(s,t;r,\beta) = 2\eta^3 \frac{t^{\alpha+1}}{\Gamma(\alpha+2)} + 4\delta\eta^2 \frac{t^\alpha}{\Gamma(\alpha+1)}, \tag{11.89}$$

$$\tilde{v}_2(s,t;r,\beta) = 0, \tag{11.90}$$

$$\tilde{v}_n(s,t;r,\beta) = 0, \quad \text{for } n \geq 2, \tag{11.91}$$

and finally one may write

$$\tilde{v}(s,t;r,\beta) = \eta s^2 t + \delta s^2 + 2\eta^3 \frac{t^{\alpha+1}}{\Gamma(\alpha+2)} + 4\delta\eta^2 \frac{t^\alpha}{\Gamma(\alpha+1)}, \tag{11.92}$$

Lower and upper bound of the solutions, respectively, are

$$\underline{v}(s,t;r,0) = \left(6 - 0.2\sqrt{-2\log_e r}\right)s^2 t + \left(1 - 0.2\sqrt{-2\log_e r}\right)s^2$$

$$+ 2\left(6 - 0.2\sqrt{-2\log_e r}\right)^3 \frac{t^{\alpha+1}}{\Gamma(\alpha+2)}$$

$$+ 4\left(1 - 0.2\sqrt{-2\log_e r}\right)\left(6 - 0.2\sqrt{-2\log_e r}\right)^2 \frac{t^{\alpha}}{\Gamma(\alpha+1)},$$

$$(11.93)$$

and

$$\overline{v}(s,t;r,1) = \left(6 + 0.2\sqrt{-2\log_e r}\right)s^2 t + \left(1 + 0.2\sqrt{-2\log_e r}\right)s^2$$

$$+ 2\left(6 + 0.2\sqrt{-2\log_e r}\right)^3 \frac{t^{\alpha+1}}{\Gamma(\alpha+2)}$$

$$+ 4\left(1 + 0.2\sqrt{-2\log_e r}\right)\left(6 + 0.2\sqrt{-2\log_e r}\right)^2 \frac{t^{\alpha}}{\Gamma(\alpha+1)}.$$

$$(11.94)$$

11.4 NUMERICAL RESULTS FOR FUZZY FRACTIONAL VIBRATION EQUATION FOR LARGE MEMBRANE

In this section, numerical solution of the aforementioned problem using HPM and ADM methods has been presented. It is a gigantic task to include here all the results with respect to various parameters and initial conditions involved in the corresponding fuzzy fractional differential equation. So, some particular values of the parameters are taken into consideration to compute the results with the discussed cases. Obtained results by the present analysis are compared with the existing solution of Yildirim et al. (2010) in special cases for $\alpha = 2$ and $r = 1$ to show the validation of the proposed analysis. Computed results are depicted in terms of plots.

Here numerical computations have been done by truncating the infinite series (11.36), (11.37), (11.44), (11.45), (11.47), (11.48), (11.54), (11.55) and (11.61), (11.62) to a finite number ($n = 3$) of terms. Triangular fuzzy solution for particular Cases 1–5 are depicted in Figs. 11.1–11.5 by varying time t from 0 to 2 and for a particular value of radius of membrane $s = 25$ and $\alpha = 3/2$. Next, interval solutions for r-cut 0.4, 0.8, and 1 and varying t from 0 to 8 for different cases have been illustrated in Figs. 11.6–11.10, respectively, with radius of membrane, $s = 25$. One may see from these figures that the crisp result ($r = 1$) is the central line and the interval solutions are spread both sides of the crisp results. Similarly for $t = 8$ and different values of r (for all the five cases), we plot the interval solutions in Figs. 11.11–11.15. It may be worth mentioning that for all the cases present results with $r = 1$ exactly agree with the solution of Yildirim et al. (2010).

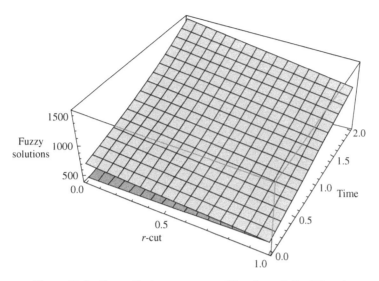

Figure 11.1 Fuzzy displacement at $s = 25$ and $\alpha = 3/2$ of Case 1

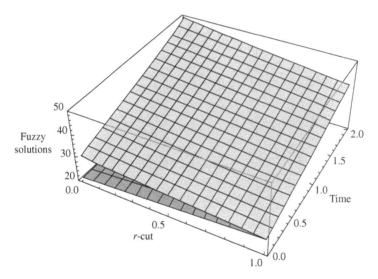

Figure 11.2 Fuzzy solution at $s = 25$ and $\alpha = 3/2$ of Case 2

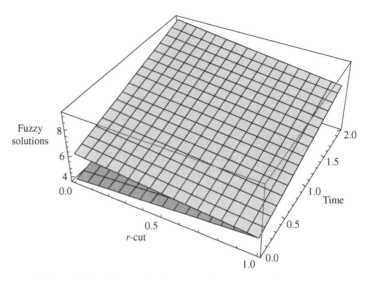

Figure 11.3 Fuzzy solution at $s = 25$ and $\alpha = 3/2$ of Case 3

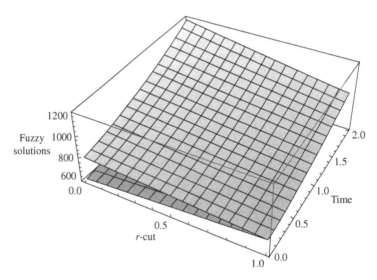

Figure 11.4 Fuzzy solution at $s = 25$ and $\alpha = 3/2$ of Case 4

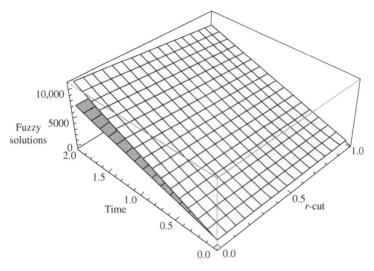

Figure 11.5 Fuzzy solution at $s = 25$ and $\alpha = 3/2$ of Case 5

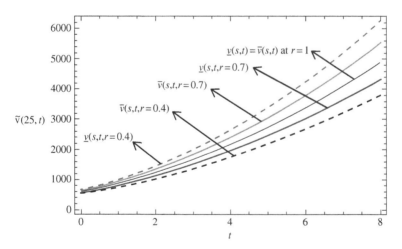

Figure 11.6 Interval solution of Case 1

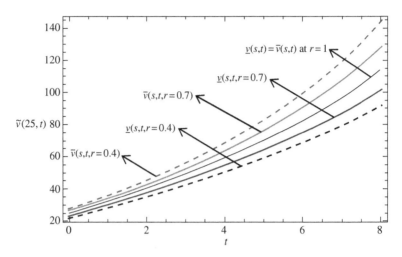

Figure 11.7   Interval solution of Case 2

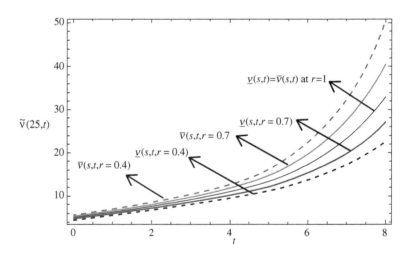

Figure 11.8   Interval solution of Case 3

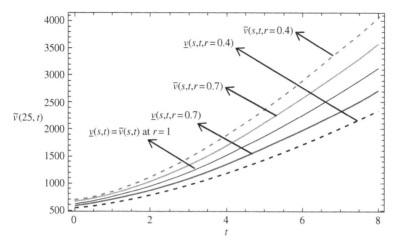

Figure 11.9 Interval solution of Case 4

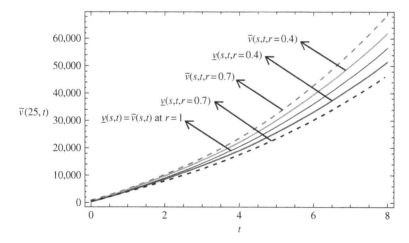

Figure 11.10 Interval solution of Case 5

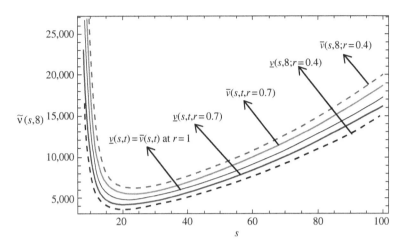

Figure 11.11   Interval solution of Case 1 at $t = 8$

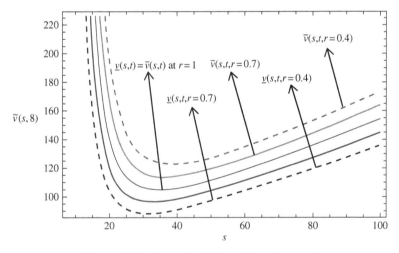

Figure 11.12   Interval solution of Case 2 at $t = 8$

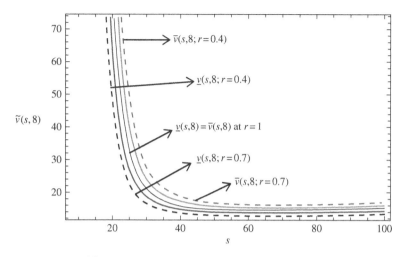

Figure 11.13   Interval solution of Case 3 at $t = 8$

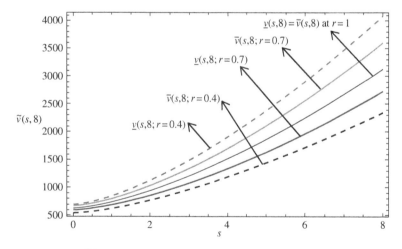

Figure 11.14   Interval solution of Case 4 at $t = 8$

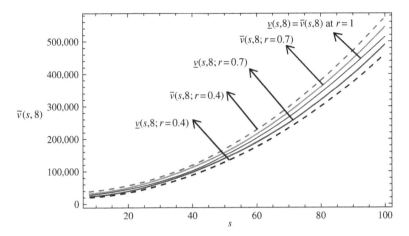

Figure 11.15   Interval solution of Case 5 at $t = 8$

Also, it is interesting to note from Figs. 11.6–11.10 that the left and right bounds of the uncertain displacement, that is $\widetilde{v}(s, t)$ (with particular values of r and s), gradually increase with increase in time. In Figs. 11.11–11.13 for particular values of r and t, the uncertain displacement first decreases and then increases with increase in radius of membrane s for Cases 1–3. But Figs. 11.14 and 11.15 for Cases 3 and 4 show that $\widetilde{v}(s, t)$ increases with increase in s. The rate of increase in uncertain displacement is faster in Case 1 than for Cases 2 and 3. Rate of increase in uncertain displacement in Case 5 is faster than that of Case 4.

Similar type of computation has been done for solution obtained by the method of ADM. Numerical computations have been done by truncating the infinite series (11.69), (11.70), (11.76), (11.77), (11.79), (11.80), (11.86), (11.87), and (11.93), (11.94) to a finite number ($n = 3$) of terms. Gaussian fuzzy solutions for particular Cases 6–10 are depicted in Figs. 11.16–11.20 by varying time t from 0 to 2 and for a particular value of radius of membrane $s = 25$. Next, interval solutions for r-cut 0.4, 0.7, and 1 and varying t from 0 to 8 for different cases have been illustrated in Figs. 11.21–11.25, respectively, with radius of membrane, $s = 25$. One may see from these figures that the crisp result ($r = 1$) is the central line and the interval solutions are spread both sides of the crisp results. Similarly for $t = 10$ and different values of r (for all the five cases), we plot the interval solutions in Figs. 11.26–11.30. It may be worth mentioning that for all the cases present results with $r = 1$ exactly agree with the solution of Yildirim et al. (2010).

It is interesting to note from Figs. 11.21–11.25 that the left and right bounds of the uncertain displacement, that is $\widetilde{v}(s, t)$ (with particular values of r and s), gradually increase with increase in time. In Figs. 11.26–11.28 for particular values of r and t, the uncertain displacement first decreases and then increases with increase in radius of membrane s for Cases 6–8. But Figs. 11.29 and 11.30 for Cases 9 and 10 show that $\widetilde{v}(s, t)$ increases with increase in s. The rate of increase in uncertain displacement is faster in Case 6 than for Cases 7 and 8. Rate of increase in uncertain displacement in Case 10 is faster than that of Case 9.

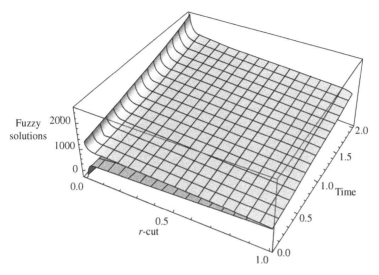

Figure 11.16 Fuzzy displacement at $s = 25$ and $\alpha = 3/2$ of Case 6

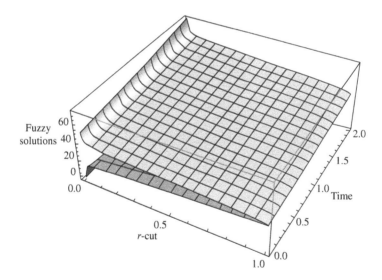

Figure 11.17 Fuzzy solution at $s = 25$ and $\alpha = 3/2$ of Case 7

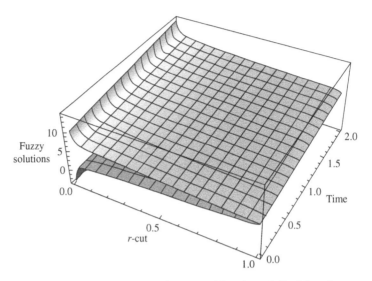

Figure 11.18   Fuzzy solution at $s = 25$ and $\alpha = 3/2$ of Case 8

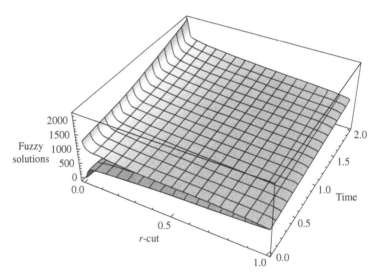

Figure 11.19   Fuzzy solution at $s = 25$ and $\alpha = 3/2$ of Case 9

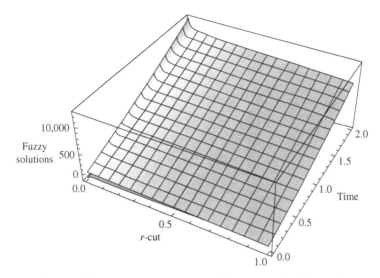

Figure 11.20   Fuzzy solution at $s = 25$ and $\alpha = 3/2$ of Case 10

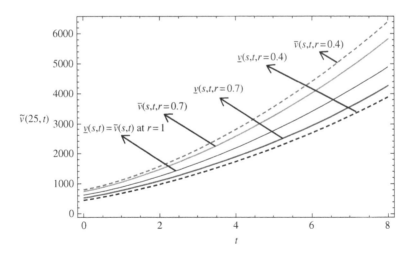

Figure 11.21   Interval solution of Case 6 for $\alpha = 3/2$

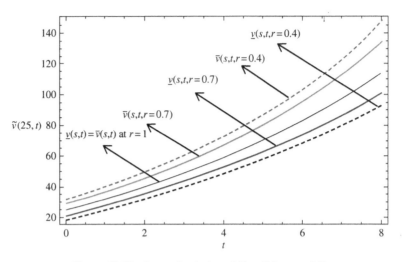

Figure 11.22   Interval solution of Case 7 for $\alpha = 3/2$

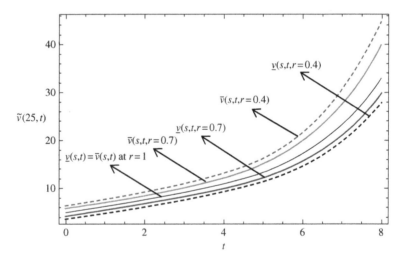

Figure 11.23   Interval solution of Case 8 for $\alpha = 3/2$

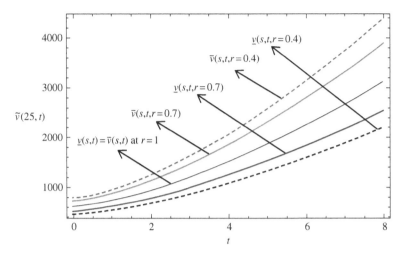

Figure 11.24 Interval solution of Case 9 for $\alpha = 3/2$

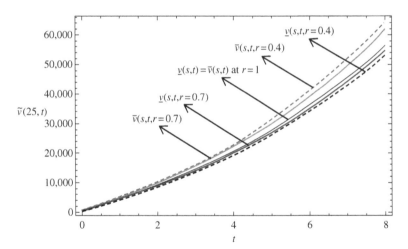

Figure 11.25 Interval solution of Case 10 for $\alpha = 3/2$

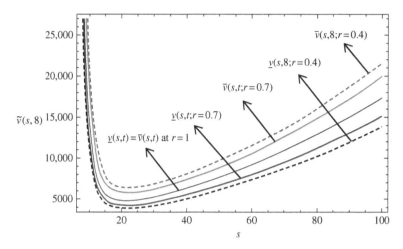

Figure 11.26   Interval solution of Case 6 at $t = 8$ for $\alpha = 3/2$

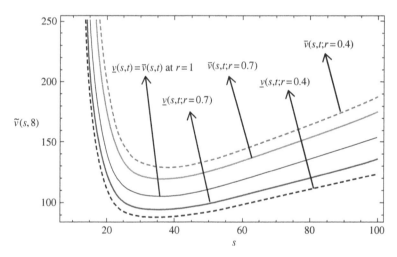

Figure 11.27   Interval solution of Case 7 at $t = 8$ for $\alpha = 3/2$

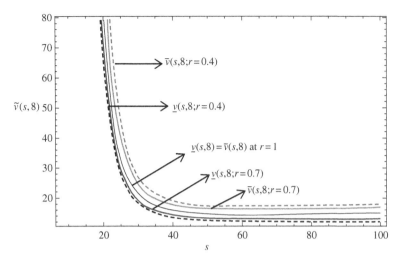

Figure 11.28   Interval solution of Case 8 at $t = 8$ for $\alpha = 3/2$

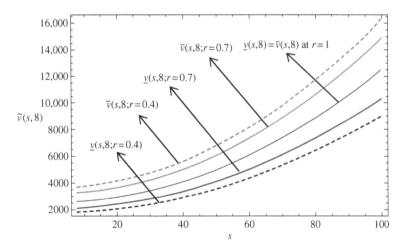

Figure 11.29   Interval solution of Case 9 at $t = 8$ for $\alpha = 3/2$

In this chapter, HPM and ADM have successfully been applied to obtain the solution of vibration equations for large membranes using double parametric form of fuzzy numbers. The proposed double parametric form approach is found to be easy and straightforward. Although the solutions by both the methods are of the form of an infinite series, it can be written in a closed form in some cases. Various case studies have been done to analyze the behavior of the membrane. The main advantages of HPM and ADM are the capability to achieve exact solution and rapid convergences with few terms.

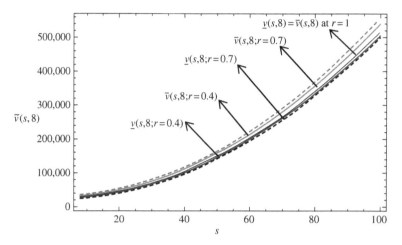

Figure 11.30   Interval solution of Case 10 at $t = 8$ for $\alpha = 3/2$

BIBLIOGRAPHY

Abbasbandy S. Numerical solutions of the integral equations: Homotopy perturbation method and Adomian's decomposition method. *Appl Math Comput* 2006;**173**:493–500.

Adomian G. A new approach to nonlinear partial differential equations. *J Math Anal Appl* 1984;**102**:420–434.

Abbaoui K, Cherruault Y. Convergence of Adomian's method applied to different equations. *Comp Math Appl* 1994;**28**:103–109.

Abbaoui K, Cherruault Y. New ideas for proving convergence of decomposition methods. *Comp Math Appl* 1995;**29**:103–108.

Adomian G. *Solving Frontier problems of Physics: The Decomposition method.* Boston: Kluwer Academic Publishers; 1994.

Behera D, Chakraverty S. Numerical solution of fractionally damped beam by homotopy perturbation method. *Cent Eur J Phys* 2013;**11**:792–798.

Biazar J, Ghazvini H. Exact solutions for nonlinear Schrödinger equations by He's homotopy perturbation method. *Phys Lett* 2007;**A366**:79–84.

Chakraverty S, Behera D. Dynamic responses of fractionally damped mechanical system using homotopy perturbation method. *Alexandria Eng J* 2013;**52**:557–562.

Chang SL, Zadeh LA. On fuzzy mapping and control. *IEEE Tran Syst Man Cybernet* 1972;**2**:30–34.

Cherruault Y. Convergence of Adomian's method. *Kyber* 1989;**18**:31–38.

Chun C. Homotopy perturbation method for a reliable analytic treatment of some evolution equations. *Z Naturforsch* 2010;**65a**:59–64.

Dubois D, Prade H. Towards fuzzy differential calculus: Part 3, differentiation. *Fuzzy Sets Syst* 1982;**8**:225–233.

Ganji DD. The applications of He's homotopy perturbation method to nonlinear equation arising in heat transfer. *Phys Lett* 2006;**A335**:337–341.

Ghanbari M. Numerical solution of fuzzy initial value problems under generalized differentiability by HPM. *Int J Ind Math* 2009;**1**:19–39.

Hanss M. *Applied Fuzzy Arithmetic: An Introduction with Engineering Applications.* Berlin: Springer-Verlag; 2005.

He JH. Homotopy perturbation technique. *Comput Methods Appl Mech Eng* 1999;**178**: 257–262.

He JH. A coupling method of homotopy technique and perturbation technique for nonlinear problems. *Int J Non-linear Mech* 2000;**35**:37–43.

Jafari M, Hosseini MM, Mohyud-Din ST, Ghovatmand M. Modified homotopy perturbation method for solving nonlinear PDAEs and its applications in nanoelectronics. *Int J Nonlinear Sci Numer Simul* 2010;**11**:1047–1057.

Jaulin L, Kieffer M, Didri OT, Walter E. *Applied Interval Analysis.* Springer; 2001.

Kaleva O. Fuzzy differential equations. *Fuzzy Sets Syst* 1987;**24**:301–317.

Kaleva O. The Cauchy problem for fuzzy differential equations. *Fuzzy Sets Syst* 1990; **35**:389–396.

Ma M, Friedman M, Kandel A. Numerical solutions of fuzzy differential equations. *Fuzzy Sets Syst* 1999;**105**:133–138.

Mohyud-Din ST, Yildirim A. An algorithm for solving the fractional vibration equation. *Comput Math Model* 2012;**23**:228–236.

Mousa M M, Kaltayev A (2010) Homotopy perturbation method for solving nonlinear Differential-Difference equations, *Z Naturforsch* **65a**: 511-517.

Nieto JJ, Rodríguez-López R, Franco D. linear first-order fuzzy differential equations. *Int J Uncertain Fuzz Knowl-Based Syst* 2006;**14**:687–709.

Noor MA, Mohyud-Din ST. Homotopy perturbation method for nonlinear higher-order boundary value problems. *Int J Nonlinear Sci Numer Simul* 2008;**9**:395–408.

Rayleigh L. *The Theory of Sound.* Dover, John William Strutt; 1945.

Ross TJ. *Fuzzy Logic with Engineering Applications.* Wiley; 2010.

Sunny ME, Kapania RK, Sultan C. Solution of nonlinear vibration problem of a prestressed membrane by Adomian decomposition. *AIAA J* 2012;**50**:1796–1800.

Seikkala S. On the fuzzy initial value problem. *Fuzzy Sets Syst* 1987;**24**:319–330.

Tapaswini S, Chakraverty S. Numerical solution of *n*-th order fuzzy linear differential equations by homotopy perturbation method. *Int J Comput Appl* 2013;**64**:5–10.

Yildirim A, Unlu C, Mohyud-Din ST. On the solution of the vibration equation by means of the homotopy perturbation method. *Appl Appl Math Special Issue* 2010;**1**:24–33.

Zimmermann HJ. *Fuzzy Set Theory and Its Application.* Boston/Dordrecht/London: Kluwer Academic Publishers; 2001.

12

FUZZY FRACTIONAL TELEGRAPH EQUATIONS

Study of fuzzy space and time fractional telegraph equations is important due to its well-known applications. There exist various investigations for the aforementioned problem where the variables and parameters are given as crisp/exact. In practice, we may not have these parameters exactly, but those may be known in some uncertain form. In this chapter, these uncertainties are considered as interval/fuzzy numbers. Again, double parametric form of fuzzy numbers has been used to solve the uncertain problem of fuzzy fractional telegraph equations along with homotopy perturbation method (HPM). The present method performs very well in terms of computational efficiency.

12.1 DOUBLE-PARAMETRIC-BASED FUZZY FRACTIONAL TELEGRAPH EQUATIONS

The fuzzy fractional telegraph equation is first converted to interval-based fuzzy fractional differential equation using single parametric form. Then by using double parametric form, interval-based fuzzy fractional telegraph differential equation has been reduced to crisp telegraph equations. Finally, HPM has been applied to solve the corresponding crisp differential equation to obtain the required solution in terms of interval/fuzzy. Let us now consider the fuzzy fractional telegraph equation

$$\frac{\partial^\alpha \widetilde{u}(x,t)}{\partial x^\alpha} = \frac{\partial^2 \widetilde{u}(x,t)}{\partial t^2} + \frac{\partial \widetilde{u}(x,t)}{\partial t} + \widetilde{u}(x,t) + \widetilde{g}(x,t), \quad t \geq 0, \ 1 < \alpha \leq 2, \quad (12.1)$$

Fuzzy Arbitrary Order System: Fuzzy Fractional Differential Equations and Applications, First Edition.
Snehashish Chakraverty, Smita Tapaswini, and Diptiranjan Behera.

subject to the fuzzy initial and boundary conditions as

$$\widetilde{u}(0,t) = \widetilde{\delta}_1 f_1(t), \quad t \geq 0, \tag{12.2}$$

$$\frac{\partial \widetilde{u}(0,t)}{\partial x} = \widetilde{\delta}_2 f_2(t), \quad t \geq 0, \tag{12.3}$$

and

$$\widetilde{u}(x,0) = \widetilde{\delta}_3 s(x), \quad 0 < x < 1, \tag{12.4}$$

where α is a parameter describing the order of the fractional space derivative. The fractional derivative is considered in the Caputo sense.

Equation (12.1) can be expressed as

$$\frac{\partial^2 \widetilde{u}(x,t)}{\partial x^2} = \frac{\partial^{2-\alpha}}{\partial x^{2-\alpha}} \left(\frac{\partial^2 \widetilde{u}(x,t)}{\partial t^2} + \frac{\partial \widetilde{u}(x,t)}{\partial t} + \widetilde{u}(x,t) + \widetilde{g}(x,t) \right), \quad t \geq 0, \ 1 < \alpha \leq 2. \tag{12.5}$$

As per single parametric form, the aforementioned fuzzy fractional telegraph equation can be written as

$$\left[\frac{\partial^2 \underline{u}(x,t)}{\partial x^2}, \frac{\partial^2 \overline{u}(x,t)}{\partial x^2} \right] = \frac{\partial^{2-\alpha}}{\partial x^{2-\alpha}} \left(\begin{array}{c} \left[\frac{\partial^2 \underline{u}(x,t)}{\partial t^2}, \frac{\partial^2 \overline{u}(x,t)}{\partial t^2} \right] + \left[\frac{\partial \underline{u}(x,t)}{\partial t}, \frac{\partial \overline{u}(x,t)}{\partial t} \right] \\ [\underline{u}(x,t), \overline{u}(x,t)] + [\underline{g}(x,t), \overline{g}(x,t)] \end{array} \right), \tag{12.6}$$

for $t \geq 0, 1 < \alpha \leq 2$, subject to fuzzy initial and boundary conditions

$$\left[\underline{u}(0,t;r), \overline{u}(0,t;r) \right] = \left[\underline{\delta}_1(r), \overline{\delta}_1(r) \right] f_1(t), \quad t \geq 0,$$

$$\left[\frac{\partial \underline{u}(0,t;r)}{\partial x}, \frac{\partial \overline{u}(0,t;r)}{\partial x} \right] = \left[\underline{\delta}_2(r), \overline{\delta}_2(r) \right] f_2(t), \quad t \geq 0,$$

$$\left[\underline{u}(x,0;r), \overline{u}(x,0;r) \right] = \left[\underline{\delta}_3(r), \overline{\delta}_3(r) \right] s(x),$$

where, $r \in [0,1], 0 < x < 1$.

Next, using the double parametric form of fuzzy numbers (as discussed in Chapter 2), Eq. (12.6) can be expressed as

$$\left(\beta \left(\frac{\partial^\alpha \overline{u}(x,t;r)}{\partial x^\alpha} - \frac{\partial^\alpha \underline{u}(x,t;r)}{\partial x^\alpha} \right) + \frac{\partial^\alpha \underline{u}(x,t;r)}{\partial x^\alpha} \right) =$$

$$\frac{\partial^{2-\alpha}}{\partial x^{2-\alpha}} \left[\begin{array}{l} \left(\left(\beta \left(\frac{\partial^2 \overline{u}(x,t;r)}{\partial t^2} - \frac{\partial^2 \underline{u}(x,t;r)}{\partial t^2} \right) + \frac{\partial^2 \underline{u}(x,t;r)}{\partial t^2} \right) \right) \\ + \left(\beta \left(\frac{\partial \overline{u}(x,t;r)}{\partial t} - \frac{\partial \underline{u}(x,t;r)}{\partial t} \right) + \frac{\partial \underline{u}(x,t;r)}{\partial t} \right) \\ + \left(\beta(\overline{u}(x,t;r) - \underline{u}(x,t;r)) + \underline{u}(x,t;r) \right) \\ + \left(\beta(\overline{g}(x,t;r) - \underline{g}(x,t;r)) + \underline{g}(x,t;r) \right) \end{array} \right], \tag{12.7}$$

subject to the initial and boundary conditions

$$\left(\beta(\overline{u}(0,t;r) - \underline{u}(0,t;r)) + \underline{u}(0,t;r)\right) = \left((\overline{\delta}_1(r) - \underline{\delta}_1(r)) + \underline{\delta}_1(r)\right)f_1(t),$$

$$\beta\left(\frac{\partial \overline{u}(0,t;r)}{\partial x} - \frac{\partial \underline{u}(0,t;r)}{\partial x}\right) + \frac{\partial \underline{u}(0,t;r)}{\partial x} = \left((\overline{\delta}_2(r) - \underline{\delta}_2(r)) + \underline{\delta}_2(r)\right)f_2(t), \quad t \geq 0,$$
$$(12.8)$$

$$\left(\beta(\overline{u}(x,0;r) - \underline{u}(x,0;r)) + \underline{u}(x,0;r)\right) = \left((\overline{\delta}_3(r) - \underline{\delta}_3(r)) + \underline{\delta}_3(r)\right)s(x), \quad 0 < x < 1,$$

where, $r, \beta \in [0,1]$.

Let us denote

$$\left(\beta\left(\frac{\partial^2 \overline{u}(x,t;r)}{\partial x^2} - \frac{\partial^2 \underline{u}(x,t;r)}{\partial x^2}\right) + \frac{\partial^2 \underline{u}(x,t;r)}{\partial x^2}\right) = \frac{\partial^2 \widetilde{u}(x,t;r,\beta)}{\partial x^2},$$

$$\left(\beta\left(\frac{\partial^2 \overline{u}(x,t;r)}{\partial t^2} - \frac{\partial^2 \underline{u}(x,t;r)}{\partial t^2}\right) + \frac{\partial^2 \underline{u}(x,t;r)}{\partial t^2}\right) = \frac{\partial^2 \widetilde{u}(x,t;r,\beta)}{\partial t^2},$$

$$\left(\beta\left(\frac{\partial \overline{u}(x,t;r)}{\partial t} - \frac{\partial \underline{u}(x,t;r)}{\partial t}\right) + \frac{\partial \underline{u}(x,t;r)}{\partial t}\right) = \frac{\partial \widetilde{u}(x,t;r,\beta)}{\partial t},$$

$$\left(\beta\left(\overline{u}(x,t;r) - \underline{u}(x,t;r)\right) + \underline{u}(x,t;r)\right) = \widetilde{u}(x,t;r,\beta),$$

$$\left(\beta\left(\overline{g}(x,t;r) - \underline{g}(x,t;r)\right) + \underline{g}(x,t;r)\right) = \widetilde{g}(x,t;r,\beta),$$

$$\left(\beta\left(\overline{u}(0,t;r) - \underline{u}(0,t;r)\right) + \underline{u}(0,t;r)\right) = \widetilde{u}(0,t;r,\beta),$$

$$\beta\left(\frac{\partial \overline{u}(0,t;r)}{\partial x} - \frac{\partial \underline{u}(0,t;r)}{\partial x}\right) + \frac{\partial \underline{u}(0,t;r)}{\partial x} = \frac{\partial \widetilde{u}(0,t;r,\beta)}{\partial x},$$

$$\left(\beta\left(\overline{u}(x,0;r) - \underline{u}(x,0;r)\right) + \underline{u}(x,0;r)\right) = \widetilde{u}(x,0;r,\beta),$$

$$\left((\overline{\delta}_1(r) - \underline{\delta}_1(r)) + \underline{\delta}_1(r)\right) = \widetilde{\delta}_1(r;\beta), \left((\overline{\delta}_2(r) - \underline{\delta}_2(r)) + \underline{\delta}_2(r)\right) = \widetilde{\delta}_2(r;\beta), \text{ and}$$

$$\left((\overline{\delta}_3(r) - \underline{\delta}_3(r)) + \underline{\delta}_3(r)\right) = \widetilde{\delta}_3(r;\beta).$$

Substituting these values in Eq. (12.7), we may write in compact form

$$\frac{\partial^2 \widetilde{u}(x,t;r,\beta)}{\partial x^2} = \frac{\partial^{2-\alpha}}{\partial x^{2-\alpha}}\left(\frac{\partial^2 \widetilde{u}(x,t;r,\beta)}{\partial t^2} + \frac{\partial \widetilde{u}(x,t;r,\beta)}{\partial t} + \widetilde{u}(x,t;r,\beta) + \widetilde{g}(x,t;r,\beta)\right)$$
$$(12.9)$$

with initial and boundary conditions

$$\widetilde{u}(0, t; r, \beta) = \widetilde{\delta}_1(r; \beta) f_1(t),$$

$$\frac{\partial \widetilde{u}(0, t; r, \beta)}{\partial x} = \widetilde{\delta}_2(r; \beta) f_2(t),$$

$$\widetilde{u}(x, 0; r, \beta) = \widetilde{\delta}_3(r; \beta) s(x).$$

Solving the corresponding crisp differential equation, one may get the solution as $\widetilde{u}(x, t; r, \beta)$ in terms of r and β. To obtain the lower and upper bounds of the solution in single parametric form, we may put $\beta = 0$ and 1, respectively. This may be represented as

$$\widetilde{u}(x, t; r, 0) = \underline{u}(x, t; r) \text{ and } \widetilde{u}(x, t; r, 1) = \overline{u}(x, t; r).$$

12.2 SOLUTIONS OF FUZZY TELEGRAPH EQUATIONS USING HOMOTOPY PERTURBATION METHOD

According to HPM, we may now construct a simple homotopy for Eq. (12.9) with an embedding parameter $p \in [0, 1]$, as follows:

$$\frac{\partial^2 \widetilde{u}(x, t; r, \beta)}{\partial x^2} = \frac{\partial^{2-\alpha}}{\partial x^{2-\alpha}} \left(\frac{\partial^2 \widetilde{u}(x, t; r, \beta)}{\partial t^2} + \frac{\partial \widetilde{u}(x, t; r, \beta)}{\partial t} + \widetilde{u}(x, t; r, \beta) + \widetilde{g}(x, t; r, \beta) \right),$$

$$(1 - p) \frac{\partial^2 \widetilde{u}(x, t; r, \beta)}{\partial x^2} + p \left(\begin{array}{c} \frac{\partial^2 \widetilde{u}(x, t; r, \beta)}{\partial x^2} - \\ \frac{\partial^{2-\alpha}}{\partial x^{2-\alpha}} \left(\begin{array}{c} \frac{\partial^2 \widetilde{u}(x, t; r, \beta)}{\partial t^2} + \frac{\partial \widetilde{u}(x, t; r, \beta)}{\partial t} \\ + \widetilde{u}(x, t; r, \beta) + \widetilde{g}(x, t; r, \beta) \end{array} \right) \end{array} \right) = 0 \quad (12.10)$$

or

$$\frac{\partial^2 \widetilde{u}(x, t; r, \beta)}{\partial x^2} + p \left(-\frac{\partial^{2-\alpha}}{\partial x^{2-\alpha}} \left(\begin{array}{c} \frac{\partial^2 \widetilde{u}(x, t; r, \beta)}{\partial t^2} + \frac{\partial \widetilde{u}(x, t; r, \beta)}{\partial t} + \\ \widetilde{u}(x, t; r, \beta) + \widetilde{g}(x, t; r, \beta) \end{array} \right) \right) = 0. \quad (12.11)$$

In the changing process from 0 to 1, for $p = 0$, Eq. (12.10) or (12.11) gives $\frac{\partial^2 \widetilde{u}(x,t;r,\beta)}{\partial x^2} = 0$, and for $p = 1$, we have the original system

$$\frac{\partial^2 \widetilde{u}(x, t; r, \beta)}{\partial x^2} - \frac{\partial^{2-\alpha}}{\partial x^{2-\alpha}} \left(\frac{\partial^2 \widetilde{u}(x, t; r, \beta)}{\partial t^2} + \frac{\partial \widetilde{u}(x, t; r, \beta)}{\partial t} + \widetilde{u}(x, t; r, \beta) + \widetilde{g}(x, t; r, \beta) \right) = 0.$$

This is called deformation in topology. Next, we can assume the solution of Eq. (12.9) as a power series expansion in p as

$$\tilde{u}(x,t;r,\beta) = \tilde{u}_0(x,t;r,\beta) + p\tilde{u}_1(x,t;r,\beta) + p^2\tilde{u}_2(x,t;r,\beta) + p^3\tilde{u}_3(x,t;r,\beta) + \cdots,$$
(12.12)

where, $\tilde{u}_i(x,t;r,\beta)$ for $i = 0, 1, 2, 3, \ldots$ are functions yet to be determined. Substituting Eq. (12.12) into Eq. (12.9) and equating the terms with the identical powers of p, we have

$$p^0 : \frac{\partial^2 \tilde{u}_0(x,t;r,\beta)}{\partial x^2} = 0,$$
(12.13)

$$p^1 : \frac{\partial^2 \tilde{u}_1(x,t;r,\beta)}{\partial x^2} - \frac{\partial^{2-\alpha}}{\partial x^{2-\alpha}} \left(\begin{array}{c} \dfrac{\partial^2 \tilde{u}_0(x,t;r,\beta)}{\partial t^2} + \dfrac{\partial \tilde{u}_0(x,t;r,\beta)}{\partial t} + \\ \tilde{u}_0(x,t;r,\beta) + \tilde{g}(x,t;r,\beta) \end{array} \right) = 0,$$
(12.14)

$$p^2 : \frac{\partial^2 \tilde{u}_2(x,t;r,\beta)}{\partial x^2} - \frac{\partial^{2-\alpha}}{\partial x^{2-\alpha}} \left(\frac{\partial^2 \tilde{u}_1(x,t;r,\beta)}{\partial t^2} + \frac{\partial \tilde{u}_1(x,t;r,\beta)}{\partial t} + \tilde{u}(x,t;r,\beta) \right) = 0,$$
(12.15)

$$p^3 : \frac{\partial^2 \tilde{u}_3(x,t;r,\beta)}{\partial x^2} - \frac{\partial^{2-\alpha}}{\partial x^{2-\alpha}} \left(\frac{\partial^2 \tilde{u}_2(x,t;r,\beta)}{\partial t^2} + \frac{\partial \tilde{u}_2(x,t;r,\beta)}{\partial t} + \tilde{u}_2(x,t;r,\beta) \right) = 0,$$
(12.16)

and so on.

Choosing initial approximation $\tilde{u}(x,0;r,\beta)$ and applying the operator $\partial^{-2}/\partial x^{-2}$ (the inverse operator of the derivative $\partial^2/\partial x^2$) on both sides of Eqs. (12.13)–(12.16), one may get the approximate solution $\tilde{u}(x,t,r,\beta) = \lim_{p \to 1} \tilde{u}(x,t;r,\beta)$, which can be expressed as

$$\tilde{u}(x,t;r,\beta) = \tilde{u}_0(x,t;r,\beta) + \tilde{u}_1(x,t;r,\beta) + \tilde{u}_2(x,t;r,\beta) + \tilde{u}_3(x,t;r,\beta) + \cdots.$$

12.3 SOLUTION BOUNDS FOR PARTICULAR CASES

In this section, we have considered fuzzy initial conditions in single parametric form as $\tilde{\delta}_1(r;\beta) = \tilde{\delta}_2(r;\beta) = \tilde{\delta}_3(r;\beta) = [0.2r + 0.8, 1.2 - 0.2r]$. Depending upon the functions $f_1(t), f_2(t), s(x)$, and $\tilde{g}(x,t;r,\beta)$, we will have different cases (Yildirim, 2010), which are discussed in the following paragraphs for finding the uncertain solution bounds.

Case 1 Here we have taken $f_1(t) = e^{-t}$, $f_2(t) = e^{-t}$, $s(x) = e^{-x}$, $\tilde{\delta}_1(r;\beta) = \tilde{\delta}_2(r;\beta) = \tilde{\delta}_3(r;\beta) = \beta(0.4 - 0.4r) + (0.2r + 0.8) = \tilde{\delta}(r;\beta)$, and $\tilde{g}(x,t;r,\beta) = 0$ in the aforementioned fuzzy initial conditions.

Hence, Eq. (12.9) will become

$$\frac{\partial^2 \widetilde{u}(x,t;r,\beta)}{\partial x^2} = \frac{\partial^{2-\alpha}}{\partial x^{2-\alpha}} \left(\frac{\partial^2 \widetilde{u}(x,t;r,\beta)}{\partial t^2} + \frac{\partial \widetilde{u}(x,t;r,\beta)}{\partial t} + \widetilde{u}(x,t;r,\beta) \right). \quad (12.17)$$

Corresponding fuzzy initial conditions in double parametric form will be

$$\widetilde{u}(0,t;r,\beta) = \widetilde{\delta}(r;\beta)e^{-t}, \quad t \geq 0,$$

$$\frac{\partial \widetilde{u}(0,t;r,\beta)}{\partial x} = \widetilde{\delta}(r;\beta)e^{-t}, \quad t \geq 0,$$

$$\widetilde{u}(x,0;r,\beta) = \widetilde{\delta}(r;\beta)e^{x}, \quad 0 < x < 1.$$

Applying HPM, one may have

$$\widetilde{u}_0(x,t;r,\beta) = \widetilde{\delta}(r;\beta)e^{-t}(1+x), \quad (12.18)$$

$$\widetilde{u}_1(x,t;r,\beta) = \widetilde{\delta}(r;\beta)e^{-t} \left(\frac{x^\alpha}{\Gamma(\alpha+1)} + \frac{x^{\alpha+1}}{\Gamma(\alpha+2)} \right), \quad (12.19)$$

$$\widetilde{u}_2(x,t;r,\beta) = \widetilde{\delta}(r;\beta)e^{-t} \left(\frac{x^{2\alpha}}{\Gamma(2\alpha+1)} + \frac{x^{2\alpha+1}}{\Gamma(2\alpha+2)} \right), \quad (12.20)$$

$$\widetilde{u}_3(x,t;r,\beta) = \widetilde{\delta}(r;\beta)e^{-t} \left(\frac{x^{3\alpha}}{\Gamma(3\alpha+1)} + \frac{x^{3\alpha+1}}{\Gamma(3\alpha+2)} \right), \quad (12.21)$$

and so on.

Therefore, the solution can be written as

$$\widetilde{u}(x,t;r,\beta) = \widetilde{\delta}(r;\beta)e^{-t} \left(\begin{array}{c} 1 + x + \dfrac{x^\alpha}{\Gamma(\alpha+1)} + \dfrac{x^{\alpha+1}}{\Gamma(\alpha+2)} + \dfrac{x^{2\alpha}}{\Gamma(2\alpha+1)} + \\[2mm] \dfrac{x^{2\alpha+1}}{\Gamma(2\alpha+2)} + \dfrac{x^{3\alpha}}{\Gamma(3\alpha+1)} + \dfrac{x^{3\alpha+1}}{\Gamma(3\alpha+2)} + \cdots \end{array} \right), \quad (12.22)$$

or

$$\widetilde{u}(x,t;r,\beta) = (\beta(0.4-0.4r) + (0.2r+0.8))e^{-t} \left(\begin{array}{c} 1 + x + \dfrac{x^\alpha}{\Gamma(\alpha+1)} + \dfrac{x^{\alpha+1}}{\Gamma(\alpha+2)} + \\[2mm] \dfrac{x^{2\alpha}}{\Gamma(2\alpha+1)} + \dfrac{x^{2\alpha+1}}{\Gamma(2\alpha+2)} + \\[2mm] \dfrac{x^{3\alpha}}{\Gamma(3\alpha+1)} + \dfrac{x^{3\alpha+1}}{\Gamma(3\alpha+2)} + \cdots \end{array} \right).$$

$$(12.23)$$

To obtain the solution bounds in single parametric form, we may put $\beta = 0$ and 1 in Eq. (12.23) for lower and upper bounds of the solution, respectively. So we get

$$\tilde{u}(x,t;r,0) = (0.2r + 0.8)e^{-t} \left(\begin{array}{l} 1 + x + \dfrac{x^\alpha}{\Gamma(\alpha + 1)} + \dfrac{x^{\alpha+1}}{\Gamma(\alpha + 2)} + \dfrac{x^{2\alpha}}{\Gamma(2\alpha + 1)} + \\[2ex] \dfrac{x^{2\alpha+1}}{\Gamma(2\alpha + 2)} + \dfrac{x^{3\alpha}}{\Gamma(3\alpha + 1)} + \dfrac{x^{3\alpha+1}}{\Gamma(3\alpha + 2)} + \cdots \end{array} \right),$$

(12.24)

and

$$\tilde{u}(x,t;r,1) = (1.2 - 0.2r)e^{-t} \left(\begin{array}{l} 1 + x + \dfrac{x^\alpha}{\Gamma(\alpha + 1)} + \dfrac{x^{\alpha+1}}{\Gamma(\alpha + 2)} + \dfrac{x^{2\alpha}}{\Gamma(2\alpha + 1)} + \\[2ex] \dfrac{x^{2\alpha+1}}{\Gamma(2\alpha + 2)} + \dfrac{x^{3\alpha}}{\Gamma(3\alpha + 1)} + \dfrac{x^{3\alpha+1}}{\Gamma(3\alpha + 2)} + \cdots \end{array} \right).$$

(12.25)

One may note that in the special case when $r = 1$, the crisp results obtained by the proposed method are exactly the same as that of the solution obtained by Yildirim (2010).

Case 2 Now we consider $f_1(t) = t, f_2(t) = 0, s(t) = x^2$,

$$\tilde{\delta}_1(r; \beta) = \tilde{\delta}_2(r; \beta) = \tilde{\delta}_3(r; \beta) = \beta(0.2\sqrt{-2\log r}) + (1 - 0.1\sqrt{-2\log r}) = \tilde{\delta}(r; \beta) \text{ and}$$
$$\tilde{g}(x,t;r,\beta) = -x^2 - t + 1.$$

Again, by applying the procedure discussed previously, one may have

$$\tilde{u}_0(x,t;r,\beta) = \tilde{\delta}(r;\beta)t, \tag{12.26}$$

$$\tilde{u}_1(x,t;r,\beta) = (\tilde{\delta}(r;\beta) + 1)\frac{x^\alpha}{\Gamma(\alpha + 1)} + \frac{tx^\alpha}{\Gamma(\alpha + 1)}\left(\tilde{\delta}(r;\beta) - 1\right) - \frac{2x^{\alpha+2}}{\Gamma(\alpha + 3)}, \tag{12.27}$$

$$\tilde{u}_2(x,t;r,\beta) = \frac{2x^{2\alpha}}{\Gamma(2\alpha + 1)} + t\frac{x^{2\alpha}}{\Gamma(2\alpha + 1)}\left(\tilde{\delta}(r;\beta) - 1\right) - \frac{2x^{2\alpha+2}}{\Gamma(2\alpha + 3)}, \tag{12.28}$$

$$\tilde{u}_3(x,t;r,\beta) = \tilde{\delta}(r;\beta)\frac{x^{3\alpha}}{\Gamma(3\alpha + 1)} + \frac{x^{3\alpha}}{\Gamma(3\alpha + 1)} + t\frac{x^{3\alpha}}{\Gamma(3\alpha + 1)}\left(\tilde{\delta}(r;\beta) - 1\right)$$
$$- \frac{2x^{3\alpha+2}}{\Gamma(3\alpha + 3)}, \tag{12.29}$$

and so on.

The solution in general form may be obtained as

$$
\tilde{u}(x,t;r,\beta) = \tilde{\delta}(r;\beta) \begin{pmatrix} t + \dfrac{x^\alpha}{\Gamma(\alpha+1)} + t\dfrac{x^\alpha}{\Gamma(\alpha+1)} + t\dfrac{x^{2\alpha}}{\Gamma(2\alpha+1)} + \dfrac{x^{3\alpha}}{\Gamma(3\alpha+1)} + \\[4mm] t\dfrac{x^{3\alpha}}{\Gamma(3\alpha+1)} \end{pmatrix}
$$

$$
+ \begin{pmatrix} \dfrac{x^\alpha}{\Gamma(\alpha+1)} - t\dfrac{x^\alpha}{\Gamma(\alpha+1)} - \dfrac{2x^{\alpha+2}}{\Gamma(\alpha+3)} + \dfrac{2x^{2\alpha}}{\Gamma(2\alpha+1)} - t\dfrac{x^{2\alpha}}{\Gamma(2\alpha+1)} - \\[4mm] \dfrac{2x^{2\alpha+2}}{\Gamma(2\alpha+3)} + \dfrac{x^{3\alpha}}{\Gamma(3\alpha+1)} - t\dfrac{x^{3\alpha}}{\Gamma(3\alpha+1)} - \dfrac{2x^{3\alpha+2}}{\Gamma(3\alpha+3)} \end{pmatrix},
$$

$$(12.30)$$

or

$$
\tilde{u}(x,t;r,\beta) = \beta\left(0.2\sqrt{-2\log r}\right) + (1 - 0.1\sqrt{-2\log r})
$$

$$
\begin{pmatrix} t + \dfrac{x^\alpha}{\Gamma(\alpha+1)} + t\dfrac{x^\alpha}{\Gamma(\alpha+1)} + t\dfrac{x^{2\alpha}}{\Gamma(2\alpha+1)} + \dfrac{x^{3\alpha}}{\Gamma(3\alpha+1)} + \\[4mm] t\dfrac{x^{3\alpha}}{\Gamma(3\alpha+1)} \end{pmatrix}
$$

$$(12.31)$$

$$
+ \begin{pmatrix} \dfrac{x^\alpha}{\Gamma(\alpha+1)} - t\dfrac{x^\alpha}{\Gamma(\alpha+1)} - \dfrac{2x^{\alpha+2}}{\Gamma(\alpha+3)} + \dfrac{2x^{2\alpha}}{\Gamma(2\alpha+1)} - t\dfrac{x^{2\alpha}}{\Gamma(2\alpha+1)} - \\[4mm] \dfrac{2x^{2\alpha+2}}{\Gamma(2\alpha+3)} + \dfrac{x^{3\alpha}}{\Gamma(3\alpha+1)} - t\dfrac{x^{3\alpha}}{\Gamma(3\alpha+1)} - \dfrac{2x^{3\alpha+2}}{\Gamma(3\alpha+3)} \end{pmatrix},
$$

Substituting $\beta = 0$ and 1 in $\tilde{v}(s,t;r,\beta)$, we get the lower and upper bounds of the fuzzy solutions, respectively, as

$$
\tilde{u}(x,t;r,0) = \left(1 - 0.1\sqrt{-2\log r}\right)
$$

$$
\begin{pmatrix} t + \dfrac{x^\alpha}{\Gamma(\alpha+1)} + t\dfrac{x^\alpha}{\Gamma(\alpha+1)} + t\dfrac{x^{2\alpha}}{\Gamma(2\alpha+1)} + \dfrac{x^{3\alpha}}{\Gamma(3\alpha+1)} + \\[4mm] t\dfrac{x^{3\alpha}}{\Gamma(3\alpha+1)} \end{pmatrix}
$$

$$(12.32)$$

$$
+ \begin{pmatrix} \dfrac{x^\alpha}{\Gamma(\alpha+1)} - t\dfrac{x^\alpha}{\Gamma(\alpha+1)} - \dfrac{2x^{\alpha+2}}{\Gamma(\alpha+3)} + \dfrac{2x^{2\alpha}}{\Gamma(2\alpha+1)} - t\dfrac{x^{2\alpha}}{\Gamma(2\alpha+1)} - \\[4mm] \dfrac{2x^{2\alpha+2}}{\Gamma(2\alpha+3)} + \dfrac{x^{3\alpha}}{\Gamma(3\alpha+1)} - t\dfrac{x^{3\alpha}}{\Gamma(3\alpha+1)} - \dfrac{2x^{3\alpha+2}}{\Gamma(3\alpha+3)} \end{pmatrix},
$$

and

$$\widetilde{u}(x, t; r, 1) = \left(1 + 0.1\sqrt{-2\log r}\right)$$

$$\left(\begin{array}{c} t + \dfrac{x^\alpha}{\Gamma(\alpha + 1)} + t\dfrac{x^\alpha}{\Gamma(\alpha + 1)} + t\dfrac{x^{2\alpha}}{\Gamma(2\alpha + 1)} + \dfrac{x^{3\alpha}}{\Gamma(3\alpha + 1)} + \\[2ex] t\dfrac{x^{3\alpha}}{\Gamma(3\alpha + 1)} \end{array}\right) \quad (12.33)$$

$$+ \left(\begin{array}{c} \dfrac{x^\alpha}{\Gamma(\alpha + 1)} - t\dfrac{x^\alpha}{\Gamma(\alpha + 1)} - \dfrac{2x^{\alpha+2}}{\Gamma(\alpha + 3)} + \dfrac{2x^{2\alpha}}{\Gamma(2\alpha + 1)} - t\dfrac{x^{2\alpha}}{\Gamma(2\alpha + 1)} - \\[2ex] \dfrac{2x^{2\alpha+2}}{\Gamma(2\alpha + 3)} + \dfrac{x^{3\alpha}}{\Gamma(3\alpha + 1)} - t\dfrac{x^{3\alpha}}{\Gamma(3\alpha + 1)} - \dfrac{2x^{3\alpha+2}}{\Gamma(3\alpha + 3)} \end{array}\right),$$

Solution obtained by the proposed method for $r = 1$ is again found to be exactly the same as that of (crisp result) (Yildirim, 2010).

12.4 NUMERICAL RESULTS FOR FUZZY FRACTIONAL TELEGRAPH EQUATIONS

In this section, we present numerical solution of fuzzy fractional telegraph equations with different $\widetilde{g}(x, t; r, \beta)$ and fuzzy initial conditions are computed by the method

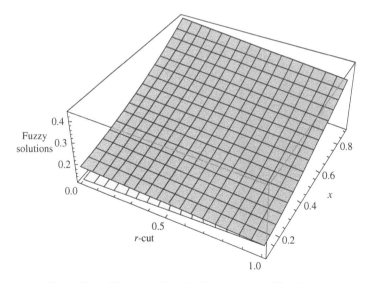

Figure 12.1 Fuzzy solution for Case 1 at $\alpha = 1.75$ and $t = 2$

of HPM. It is a gigantic task to include here all the results with respect to various parameters and initial conditions involved in the corresponding fuzzy fractional differential equation. So, some particular values of the parameters are taken to compute the results with the aforementioned cases. Obtained results by the present analysis are compared with the existing solution of Yildirim (2010) in special cases to show the validation of the proposed analysis. Computed results are depicted in terms of plots.

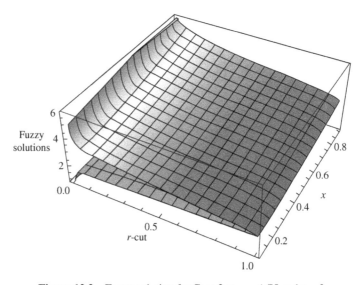

Figure 12.2   Fuzzy solution for Case 2 at $\alpha = 1.75$ and $t = 2$

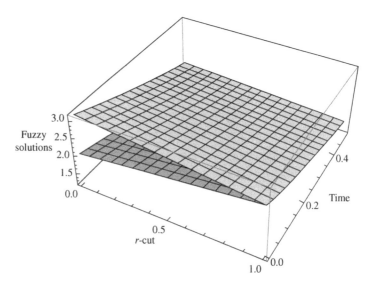

Figure 12.3   Fuzzy solution for Case 1 at $\alpha = 1.5$ and $x = 0.8$

Triangular and Gaussian fuzzy solutions are depicted in Figs. 12.1 and 12.2, respectively, by varying x from 0.1 to 0.9 and for a particular value $\alpha = 1.75$ and $t = 2$ for both the cases. Similarly, the solutions are depicted in Figs. 12.3 and 12.4, by varying t from 0 to 0.5 with the constant values $\alpha = 1.5$ and $x = 0.8$. Next, interval solutions for both the cases r-cut 0.4, 0.7, and 1, $\alpha = 1.75$ and varying x from 0.1 to 0.9 have been depicted in Figs. 12.5 and 12.6 with $t = 2$. Also, interval solutions for

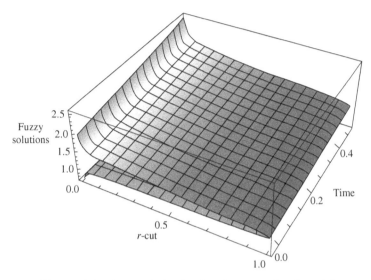

Figure 12.4   Fuzzy solution for Case 2 at $\alpha = 1.5$ and $x = 0.8$

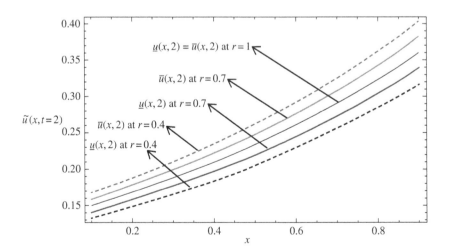

Figure 12.5   Interval solution for Case 1 at $\alpha = 1.75$ and $t = 2$

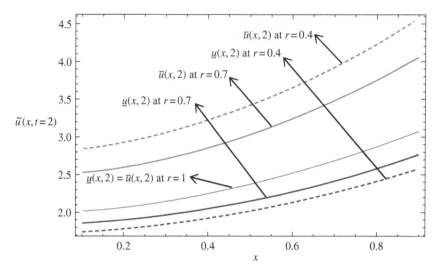

Figure 12.6   Interval solution for Case 2 at $\alpha = 1.75$ and $t = 2$

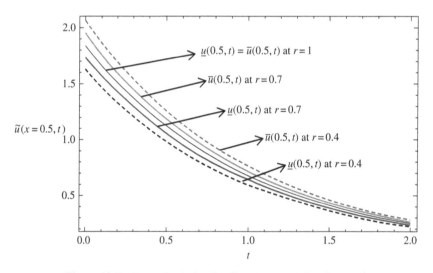

Figure 12.7   Interval solution for Case 1 at $\alpha = 1.5$ and $x = 0.5$

both the cases have been plotted in Figs. 12.7 and 12.8 with the parameters $\alpha = 1.5$, r-cut 0.4, 0.7, and 1 and varying x from 0.1 to 0.9.

Also, it is interesting to note from Fig. 12.9 that the left and right bounds of the uncertain $\widetilde{u}(x, t)$ (with particular values of r, x and α) gradually decrease with increase in time, and as shown in Fig. 12.10, the uncertain $\widetilde{u}(x, t)$ (with particular

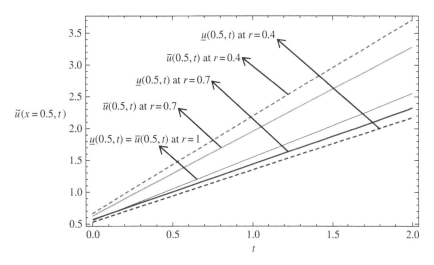

Figure 12.8   Interval solution for Case 2 at $\alpha = 1.5$ and $x = 0.5$

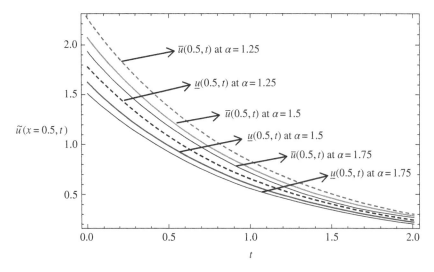

Figure 12.9   Interval solution for Case 1 at $r = 0.4$ and $x = 0.5$

values of r, t and α) gradually increases with increase in the value of x. For Case 2, as shown in Figs. 12.11 and 12.12, the $\widetilde{u}(x, t)$ increases with increase in time and x, respectively.

In this chapter, double parametric form of fuzzy numbers has successfully been applied to the solution of fuzzy fractional telegraph equation using HPM. Double

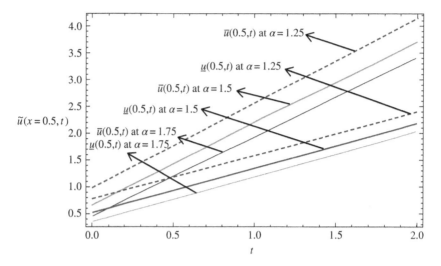

Figure 12.10   Interval solution for Case 1 at $r = 0.6$ and $t = 2$

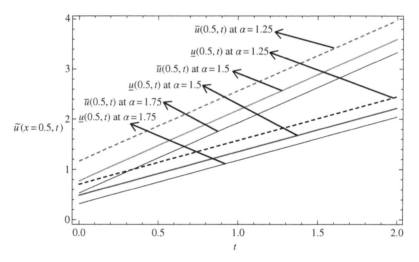

Figure 12.11   Interval solution for Case 2 at $r = 0.4$ and $x = 0.5$

parametric form approach is found to be easy and straightforward. Performance of the method is shown by using triangular and Gaussian fuzzy numbers. It is interesting to note that the lower bound is equal to the upper bound solution for $r = 1$. Although the solution by HPM is of the form of an infinite series, it can be written in a closed form. The main advantage of HPM is the capability to achieve exact solution and rapid convergence with few terms.

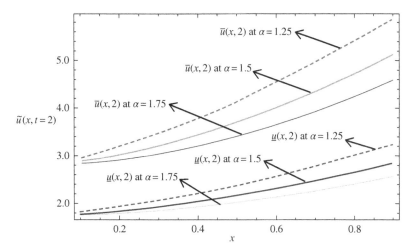

Figure 12.12 Interval solution for Case 2 at $r = 0.6$ and $t = 2$

BIBLIOGRAPHY

Abdulaziz O, Hashim I, Momani S. Application of homotopy-perturbation method to fractional IVPs. *J Comput Appl Math* 2008;**216**:574–584.

Balci MA, Yildirim A. Analysis of fractional nonlinear differential equations using the homotopy perturbation method. *Z Naturforsch* 2011;**A66**:87–92.

Behera D, Chakraverty S. Numerical solution of fractionally damped beam by homotopy perturbation method. *Cent Eur J Phys* 2013;**11**:792–798.

Biazar J, Ghazvini H. Exact solutions for nonlinear Schrödinger equations by He's homotopy perturbation method. *Phys Lett* 2007;**A366**:79–84.

Chakraverty S, Behera D. Dynamic responses of fractionally damped mechanical system using homotopy perturbation method. *Alexandria Eng J* 2013;**52**:557–562.

Chalco-Cano Y, Roman-Flores H. On new solutions of fuzzy differential equations. *Chaos Solitons Fractals* 2008;**38**:112–119.

Chang SL, Zadeh LA. On fuzzy mapping and control. *IEEE Trans Syst Man Cybernet* 1972;**2**:30–34.

Chun C. Homotopy perturbation method for a reliable analytic treatment of some evolution equations. *Z Naturforsch* 2010;**A65**:59–64.

Cui FGM. Homotopy perturbation–reproducing kernel method for nonlinear systems of second order boundary value problems. *J Comput Appl Math* 2011;**235**:2405–2411.

Dubois D, Prade H. Towards fuzzy differential calculus: Part 3, differentiation. *Fuzzy Sets Syst* 1982;**8**:225–233.

Ganji DD. The applications of He's homotopy perturbation method to nonlinear equation arising in heat transfer. *Phys Lett* 2006;**A335**:337–341.

He JH. Homotopy perturbation technique. *Comput Methods Appl Mech Eng* 1999;**178**:257–262.

He JH. A coupling method of homotopy technique and perturbation technique for nonlinear problems. *Int J Nonlinear Mech* 2000;**35**:37–43.

Hanss M. *Applied Fuzzy Arithmetic: An Introduction with engineering applications*. Berlin: Springer-Verlag; 2005.

Jaulin L, Kieffer M, Didri OT, Walter E. *Applied Interval Analysis*. Springer; 2001.

Kaleva O. Fuzzy differential equations. *Fuzzy Sets Syst* 1987;**24**:301–317.

Kaleva O. The Cauchy problem for fuzzy differential equations. *Fuzzy Sets Syst* 1990; **35**:389–396.

Khastan A, Nieto JJ, Rodrıguez- Lopez R. Variation of constant formula for first order fuzzy differential equations. *Fuzzy Sets Syst* 2011;**177**:20–33.

Lu J-H, Zheng CL. Approximate solution of generalized Ginzburg–Landau–Higgs system via homotopy perturbation method. *Z Naturforsch* 2010;**A65**:301–304.

Ma M, Friedman M, Kandel A. Numerical solutions of fuzzy differential equations. *Fuzzy Sets Syst* 1999;**105**:133–138.

Matinfar M, Saeidy M. Application of homotopy perturbation method for fuzzy integral equations. *J Math Comput Sci* 2010;**1**:377–385.

Mikaeilvand N, Khakrangin S. Solving fuzzy partial differential equations by fuzzy two-dimensional differential transform method. *Neural Comput Appl* 2012;**21**: S307–S312.

Mousa MM, Kaltayev A. Homotopy perturbation method for solving nonlinear differential-difference equations. *Z Naturforsch* 2010;**A65**:511–517.

Nieto JJ, Rodríguez-López R, Franco D. Linear first-order fuzzy differential equations. *Int J Unc Fuzz Knowl Based Syst* 2006;**14**:687–709.

Ozis T, Akçı C. Periodic solutions for certain non-smooth oscillators by iterated homotopy perturbation method combined with modified Lindstedt–Poincare technique. *Meccanica* 2011;**46**:341–347.

Palligkinis S, Papageorgiou G, Famelis I. Runge-Kutta methods for fuzzy differential equations. *Appl Math Comput* 2009;**209**:97–105.

Ross TJ. *Fuzzy Logic with Engineering Applications*. Wiley Student Edition; 2007.

Seikkala S. On the fuzzy initial value problem. *Fuzzy Sets Syst* 1987;**24**:319–330.

Tapaswini S, Chakraverty S. A new approach to fuzzy initial value problem by improved Euler method. *Int J Fuzzy Inf Eng* 2012;**4**:293–312.

Tapaswini S, Chakraverty S. Numerical solution of *n*-th order fuzzy linear differential equations by homotopy perturbation method. *Int J Comput Appl* 2013;**64**:5–10.

Tapaswini S, Chakraverty S. Euler based new solution method for fuzzy initial value problems. *Int J Artif Intell Soft Comput* 2014;**4**:58–79.

Yildirim A. He's homotopy perturbation method for solving the space- and time- fractional telegraph equations. *Int J Comput Math* 2010;**87**:2998–3006.

Zimmermann HJ. *Fuzzy Set Theory and Its Application*. Boston/Dordrecht/London: Kluwer Academic Publishers; 2001.

13

FUZZY FOKKER–PLANCK EQUATION WITH SPACE AND TIME FRACTIONAL DERIVATIVES

This chapter analyzed the solution of fuzzy fractional Fokker–Planck equation with initial conditions as triangular fuzzy number. In the solution process, homotopy perturbation method (HPM) and Adomian decomposition method (ADM) have been used along with the concept of double parametric form of fuzzy numbers.

13.1 FUZZY FRACTIONAL FOKKER–PLANCK EQUATION WITH SPACE AND TIME FRACTIONAL DERIVATIVES

Similarly to the previous chapters, fuzzy fractional Fokker–Planck equation with space and time fractional derivatives has been first converted to an interval-based fuzzy fractional Fokker–Planck equation using a single parametric form of fuzzy numbers. Then the interval-based differential equation is reduced to a crisp differential equation by using double parametric form of fuzzy numbers (similarly to the procedure given in the previous chapters). Then HPM and ADM have been applied to obtain the solution in double parametric form.

Let us consider the following fuzzy fractional Fokker–Planck equation:

$$\frac{\partial^\alpha \widetilde{u}}{\partial t^\alpha} = \left[-\frac{\partial^\gamma}{\partial x^\gamma} A(x) + \frac{\partial^{2\gamma}}{\partial x^{2\gamma}} B(x) \right] \widetilde{u}(x,t) \quad t > 0, \ x > 0, \ 0 < \alpha, \ \gamma \le 1, \quad (13.1)$$

with fuzzy initial condition $\widetilde{u}(x,0) = \widetilde{\delta}x$, where $\partial^\alpha / \partial t^\alpha$, $\partial^\gamma / \partial x^\gamma$, and $\partial^{2\gamma} / \partial x^{2\lambda}$ are the Caputo derivatives of order α and γ, $B(x) > 0$ is the diffusion coefficient, and $A(x)$

Fuzzy Arbitrary Order System: Fuzzy Fractional Differential Equations and Applications, First Edition.
Snehashish Chakraverty, Smita Tapaswini, and Diptiranjan Behera.
© 2016 John Wiley & Sons, Inc. Published 2016 by John Wiley & Sons, Inc.

is the drift coefficient. The drift and diffusion coefficients may also depend on time. The function $\tilde{u}(x,t)$ is assumed to be a casual uncertain function of time and space, that is, vanishing for $t < 0$ and $x < 0$.

The aforementioned fuzzy fractional Fokker–Planck equation (13.1) can be written in single parametric form as

$$
\left[\frac{\partial^\alpha \underline{u}(x,t;r)}{\partial t^\alpha}, \frac{\partial^\alpha \overline{u}(x,t;r)}{\partial t^\alpha}\right] = -\left[\frac{\partial^\gamma \underline{u}(x,t;r)}{\partial x^\gamma}A(x), \frac{\partial^\gamma \overline{u}(x,t;r)}{\partial x^\gamma}A(x)\right]
$$

$$
+ \left[\frac{\partial^{2\gamma} \underline{u}(x,t;r)}{\partial x^{2\gamma}}B(x), \frac{\partial^{2\gamma} \overline{u}(x,t;r)}{\partial x^{2\gamma}}B(x)\right],
$$

$$
t > 0, \ x > 0, \ 0 < \alpha, \ \gamma \leq 1, \tag{13.2}
$$

subject to single parametric fuzzy initial condition

$$
[\underline{u}(x,0;r), \overline{u}(x,0;r)] = [\underline{\delta}(r), \overline{\delta}(r)]x.
$$

Using the double parametric form (as discussed in Chapter 1), Eq. (13.2) can be expressed as

$$
\left\{\beta\left(\frac{\partial^\alpha \overline{u}(x,t;r)}{\partial t^\alpha} - \frac{\partial^\alpha \underline{u}(x,t;r)}{\partial t^\alpha}\right) + \frac{\partial^\alpha \underline{u}(x,t;r)}{\partial t^\alpha}\right\}
$$

$$
= -A(x)\left\{\beta\left(\frac{\partial^\gamma \overline{u}(x,t;r)}{\partial x^\gamma} - \frac{\partial^\gamma \underline{u}(x,t;r)}{\partial x^\gamma}\right) + \frac{\partial^\gamma \underline{u}(x,t;r)}{\partial x^\gamma}\right\}
$$

$$
+ B(x)\left\{\beta\left(\frac{\partial^{2\gamma} \overline{u}(x,t;r)}{\partial x^{2\gamma}} - \frac{\partial^{2\gamma} \underline{u}(x,t;r)}{\partial x^{2\gamma}}\right) + \frac{\partial^{2\gamma} \underline{u}(x,t;r)}{\partial x^{2\gamma}}\right\} \tag{13.3}
$$

subject to the fuzzy initial condition

$$
\{\beta(\overline{u}(x,0;r) - \underline{u}(x,0;r)) + \underline{u}(x,0;r)\} = \{\beta(\overline{\delta}(r) - \underline{\delta}(r)) + \underline{\delta}(r)\}x, \text{ where } r, \beta \in [0,1].
$$

Let us now denote

$$
\left\{\beta\left(\frac{\partial^\alpha \overline{u}(x,t;r)}{\partial t^\alpha} - \frac{\partial^\alpha \underline{u}(x,t;r)}{\partial t^\alpha}\right) + \frac{\partial^\alpha \underline{u}(x,t;r)}{\partial t^\alpha}\right\} = \frac{\partial^\alpha \tilde{u}(x,t;r,\beta)}{\partial t^\alpha},
$$

$$
\left\{\beta\left(\frac{\partial^\gamma \overline{u}(x,t;r)}{\partial x^\gamma} - \frac{\partial^\gamma \underline{u}(x,t;r)}{\partial x^\gamma}\right) + \frac{\partial^\gamma \underline{u}(x,t;r)}{\partial x^\gamma}\right\} = \frac{\partial^\gamma \tilde{u}(x,t;r,\beta)}{\partial x^\gamma},
$$

$$
\left\{\beta\left(\frac{\partial^{2\gamma} \overline{u}(x,t;r)}{\partial x^{2\gamma}} - \frac{\partial^{2\gamma} \underline{u}(x,t;r)}{\partial x^{2\gamma}}\right) + \frac{\partial^{2\gamma} \underline{u}(x,t;r)}{\partial x^{2\gamma}}\right\} = \frac{\partial^{2\gamma} \tilde{u}(x,t;r,\beta)}{\partial x^{2\gamma}},
$$

$$\{\beta(\overline{u}(x,0;r) - \underline{u}(x,0;r)) + \underline{u}(x,0;r)\}$$

$$= \widetilde{u}(x,0;r,\beta) \quad \text{and} \quad \{\beta(\overline{\delta}(r) - \underline{\delta}(r)) + \underline{\delta}(r)\} = \widetilde{\delta}(r,\beta).$$

Substituting these in Eq. (13.3), we get

$$\frac{\partial^\alpha \widetilde{u}(x,t;r,\beta)}{\partial t^\alpha} = -A(x)\frac{\partial^\gamma \widetilde{u}(x,t;r,\beta)}{\partial x^\gamma} + B(x)\frac{\partial^{2\gamma} \widetilde{u}(x,t;r,\beta)}{\partial x^{2\gamma}} \qquad (13.4)$$

with initial condition

$$\widetilde{u}(x,0;r,\beta) = \widetilde{\delta}(r,\beta)x.$$

Solving the corresponding crisp differential equation (Eq. (13.4)), one may get the solution as $\widetilde{u}(x,t;r,\beta)$. The lower and upper bounds of the solution in single parametric form have been obtained by putting $\beta = 0$ and 1, respectively, which may be represented as

$$\widetilde{u}(x,t;r,0) = \underline{u}(x,t;r) \quad \text{and} \quad \widetilde{u}(x,t;r,1) = \overline{u}(x,t;r).$$

Now HPM and ADM have been applied to solve Eq. (13.4).

13.2   DOUBLE-PARAMETRIC-BASED SOLUTION OF UNCERTAIN FRACTIONAL FOKKER–PLANCK EQUATION

As mentioned earlier, two methods, namely HPM and ADM, have been used to solve the Fokker–Planck uncertain fractional equations.

13.2.1   Solution by HPM

In this section, HPM has been applied to solve Eq. (13.4). A simple homotopy is constructed for an embedding parameter $p \in [0, 1]$ as follows:

$$(1 - p)\frac{\partial^\alpha \widetilde{u}(x,t;r,\beta)}{\partial t^\alpha}$$

$$+ p\left(\frac{\partial^\alpha \widetilde{u}(x,t;r,\beta)}{\partial t^\alpha} + A(x)\frac{\partial^\gamma \widetilde{u}(x,t;r,\beta)}{\partial x^\gamma} - B(x)\frac{\partial^{2\gamma} \widetilde{u}(x,t;r,\beta)}{\partial x^{2\gamma}}\right) = 0 \quad (13.5)$$

or

$$\frac{\partial^\alpha \widetilde{u}(x,t;r,\beta)}{\partial t^\alpha} + p\left(A(x)\frac{\partial^\gamma \widetilde{u}(x,t;r,\beta)}{\partial x^\gamma} - B(x)\frac{\partial^{2\gamma} \widetilde{u}(x,t;r,\beta)}{\partial x^{2\gamma}}\right) = 0. \qquad (13.6)$$

In the changing process from 0 to 1, for $p = 0$, Eq. (13.5) or (13.6) gives $\partial^\alpha \widetilde{u}(x,t;r,\beta)/\partial t^\alpha = 0$ and for $p = 1$, one may have the original system

$$\frac{\partial^\alpha \widetilde{u}(x,t;r,\beta)}{\partial t^\alpha} + A(x)\frac{\partial^\gamma \widetilde{u}(x,t;r,\beta)}{\partial x^\gamma} - B(x)\frac{\partial^{2\gamma} \widetilde{u}(x,t;r,\beta)}{\partial x^{2\gamma}} = 0.$$

This is called deformation in topology.

$$\frac{\partial^\alpha \tilde{u}(x,t;r,\beta)}{\partial t^\alpha} \quad \text{and} \quad A(x)\frac{\partial^\gamma \tilde{u}(x,t;r,\beta)}{\partial x^\gamma} - B(x)\frac{\partial^{2\gamma} \tilde{u}(x,t;r,\beta)}{\partial x^{2\gamma}}$$

are called homotopic. We can assume the solution of Eq. (13.1) or (13.2) as a power series expansion in p as

$$\tilde{u}(x,t;r,\beta) = \tilde{u}_0(x,t;r,\beta) + p\tilde{u}_1(x,t;r,\beta) + p^2\tilde{u}_2(x,t;r,\beta) + p^3\tilde{u}_3(x,t;r,\beta) + \cdots$$
$$(13.7)$$

where $\tilde{u}_i(x,t;r,\beta)$ for $i = 0,1,2,3,\ldots$ are functions yet to be determined. Substituting Eq. (13.7) into Eq. (13.5) or (13.6) and equating the terms with the identical powers of p, we have

$$p^0 : \quad \frac{\partial^\alpha \tilde{u}_0(x,t;r,\beta)}{\partial t^\alpha} = 0, \tag{13.8}$$

$$p^1 : \quad \frac{\partial^\alpha \tilde{u}_1(x,t;r,\beta)}{\partial t^\alpha} + A(x)\frac{\partial^\gamma \tilde{u}_0(x,t;r,\beta)}{\partial x^\gamma} - B(x)\frac{\partial^{2\gamma} \tilde{u}_0(x,t;r,\beta)}{\partial x^{2\gamma}} = 0, \tag{13.9}$$

$$p^2 : \quad \frac{\partial^\alpha \tilde{u}_2(x,t;r,\beta)}{\partial t^\alpha} + A(x)\frac{\partial^\gamma \tilde{u}_1(x,t;r,\beta)}{\partial x^\gamma} - B(x)\frac{\partial^{2\gamma} \tilde{u}_1(x,t;r,\beta)}{\partial x^{2\gamma}} = 0, \tag{13.10}$$

$$p^3 : \quad \frac{\partial^\alpha \tilde{u}_3(x,t;r,\beta)}{\partial t^\alpha} + A(x)\frac{\partial^\gamma \tilde{u}_2(x,t;r,\beta)}{\partial x^\gamma} - B(x)\frac{\partial^{2\gamma} \tilde{u}_2(x,t;r,\beta)}{\partial x^{2\gamma}} = 0, \tag{13.11}$$

and so on.

Choosing initial approximation $\tilde{u}(x,0;r,\beta)$ and applying the operator J^α (the inverse operator of Caputo derivative of order α) on both sides of Eqs. (13.8)–(13.11), one may get the approximate solution $\tilde{u}(x,t;r,\beta) = \lim_{p \to 1} \tilde{u}(x,t;r,\beta)$, which can be expressed as

$$\tilde{u}(x,t;r,\beta) = \tilde{u}_0(x,t;r,\beta) + \tilde{u}_1(x,t;r,\beta) + \tilde{u}_2(x,t;r,\beta) + \tilde{u}_3(x,t;r,\beta) + \cdots \quad (13.12)$$

13.2.2   Solution By ADM

Let us now consider Eq. (13.4) as

$$L_t^\alpha \tilde{u}(x,t;r,\beta) = -A(x)L_x^\gamma \tilde{u}(x,t;r,\beta) + B(x)L_x^{2\gamma}\tilde{u}(x,t;r,\beta), \tag{13.13}$$

where $L_t^\alpha = \partial^\alpha/\partial t^\alpha$, $L_x^\gamma = \partial^\gamma/\partial x^\gamma$, and $L_x^{2\gamma} = \partial^{2\gamma}/\partial x^{2\gamma}$.

Applying the operator $L_t^{-\alpha}$ (which is the inverse operator of L_t^α) on both sides of Eq. (13.13), obtained equivalent expression is given as

$$L_t^{-\alpha}L_t^\alpha \tilde{u}(x,t;r,\beta) = L_t^{-\alpha}(-A(x)L_x^\gamma \tilde{u}(x,t;r,\beta) + B(x)L_x^{2\gamma}\tilde{u}(x,t;r,\beta)). \tag{13.14}$$

Now, we have

$$L_t^{-\alpha} L_t^{\alpha} \widetilde{u}(x, t; r, \beta) = \widetilde{u}(x, 0; r, \beta).$$

Then Eq. (13.14) becomes

$$\widetilde{u}(x, t; r, \beta) = \widetilde{u}(x, 0; r, \beta) + L_t^{-\alpha}(-A(x)L_x^{\gamma}\widetilde{u}(x, t; r, \beta) + B(x)L_x^{2\gamma}\widetilde{u}(x, t; r, \beta)). \quad (13.15)$$

According to ADM (Adomian, 1984, 1994), we assume an infinite series solution for unknown function $\widetilde{u}(x, t; r, \beta)$ as

$$\widetilde{u}(x, t; r, \beta) = \sum_{n=0}^{\infty} \widetilde{u}_n(x, t; r, \beta), \quad (13.16)$$

with $\widetilde{u}_0(x, t; r, \beta) = \widetilde{u}(x, 0; r, \beta)$ and the components $\widetilde{u}_n(x, t; r, \beta)$ where $n > 0$ are usually determined by

$$\widetilde{u}_1(x, t; r, \beta) = L_t^{-\alpha}(-A(x)L_x^{\gamma}\widetilde{u}_0(x, t; r, \beta) + B(x)L_x^{2\gamma}\widetilde{u}_0(x, t; r, \beta)),$$

$$\widetilde{u}_2(x, t; r, \beta) = L_t^{-\alpha}(-A(x)L_x^{\gamma}\widetilde{u}_1(x, t; r, \beta) + B(x)L_x^{2\gamma}\widetilde{u}_1(x, t; r, \beta)),$$

$$\widetilde{u}_3(x, t; r, \beta) = L_t^{-\alpha}(-A(x)L_x^{\gamma}\widetilde{u}_2(x, t; r, \beta) + B(x)L_x^{2\gamma}\widetilde{u}_2(x, t; r, \beta)),$$

and so on.

Now substituting these terms in Eq. (13.16), one may get the approximate solution of Eq. (13.4) as follows:

$$\widetilde{u}(x, t; r, \beta) = \widetilde{u}_0(x, t; r, \beta) + \widetilde{u}_1(x, t; r, \beta) + \widetilde{u}_2(x, t; r, \beta) + \widetilde{u}_3(x, t; r, \beta) + \cdots. \quad (13.17)$$

As mentioned earlier, the series obtained by ADM converges very rapidly and only few terms are required to get the approximate solutions. The proof may be found in Abbaoui and Cherruault (1994, 1995), Cherruault (1989), and Himoun et al. (1999).

13.3   CASE STUDIES USING HPM AND ADM

13.3.1   Using HPM

In this section, two different Cases 1 and 2 have been considered as follows.

Case 1   Let us consider $\alpha = 1$, $A(x) = x$, and $B(x) = x^2/2$ with triangular fuzzy initial condition in parametric form as $\widetilde{u}(x, 0; r) = [0.1r + 0.9, 1.1 - 0.1r]x$. Hence, Eq. (13.1) will become

$$\frac{\partial \widetilde{u}}{\partial t} = \left[-\frac{\partial^{\gamma}}{\partial x^{\gamma}} x + \frac{\partial^{2\gamma}}{\partial x^{2\gamma}} \frac{x^2}{2} \right] \widetilde{u}(x, t), \quad t > 0, \quad x > 0, \quad 0 < \gamma \le 1. \quad (13.18)$$

Using double parametric form, Eq. (13.18) and the corresponding fuzzy initial condition will become

$$\frac{\partial^\alpha \tilde{u}(x,t;r,\beta)}{\partial t^\alpha} = \left[-\frac{\partial^\gamma}{\partial x^\gamma} x + \frac{\partial^{2\gamma}}{\partial x^{2\gamma}} \frac{x^2}{2} \right] \tilde{u}(x,t;r,\beta)$$

and

$$\tilde{u}(x,\ 0;r,\beta) = \beta(0.2 - 0.2r) + (0.1r + 0.9)x.$$

Applying HPM, we have

$$\tilde{u}_0(x,t;r,\beta) = \{\beta(0.2 - 0.2r) + (0.1r + 0.9)\}x, \tag{13.19}$$

$$\tilde{u}_1(x,t;r,\beta) = \{\beta(0.2 - 0.2r) + (0.1r + 0.9)\}t$$
$$\times \left[\frac{3x^{3-2\gamma}}{\Gamma(4-2\gamma)} - \frac{2x^{2-\gamma}}{\Gamma(3-\gamma)} \right], \tag{13.20}$$

$$\tilde{u}_2(x,t;r,\beta) = \{\beta(0.2 - 0.2r) + (0.1r + 0.9)\}$$
$$\times \frac{t^2}{2} \left[\begin{array}{l} \dfrac{2\Gamma(4-\gamma)}{\Gamma(3-\gamma)\Gamma(4-2\gamma)} x^{3-2\gamma} - \\[2ex] \left(\dfrac{3\Gamma(5-2\gamma)}{\Gamma(4-2\gamma)} + \dfrac{\Gamma(5-\gamma)}{\Gamma(3-\gamma)} \right) \dfrac{x^{4-3\gamma}}{\Gamma(5-3\gamma)} + \\[2ex] \dfrac{3}{2} \dfrac{\Gamma(6-2\gamma)x^{5-4\gamma}}{\Gamma(4-2\gamma)\Gamma(6-4\gamma)} \end{array} \right]. \tag{13.21}$$

and so on.

Therefore, the solution can be written as

$$\tilde{u}(x,t;r,\beta) = \{\beta(0.2 - 0.2r) + (0.1r + 0.9)\}$$
$$\times \left[\begin{array}{l} x + \left(\dfrac{3x^{3-2\gamma}}{\Gamma(4-2\gamma)} - \dfrac{2x^{2-\gamma}}{\Gamma(3-\gamma)} \right) t + \\[2ex] \left\{ \begin{array}{l} \dfrac{2\Gamma(4-\gamma)x^{3-2\gamma}}{\Gamma(3-\gamma)\Gamma(4-2\gamma)} - \\[2ex] \left(\dfrac{3\Gamma(5-2\gamma)}{\Gamma(4-2\gamma)} + \dfrac{\Gamma(5-\gamma)}{\Gamma(3-\gamma)} \right) \dfrac{x^{4-3\gamma}}{\Gamma(5-3\gamma)} \\[2ex] + \dfrac{3}{2} \dfrac{\Gamma(6-2\gamma)x^{5-4\gamma}}{\Gamma(4-2\gamma)\Gamma(6-4\gamma)} \end{array} \right\} \dfrac{t^2}{2} + \end{array} \right]. \tag{13.22}$$

To obtain the solution bound in single parametric form, we may put $\dot{\beta} = 0$ and 1 for lower and upper bounds of the solution, respectively, to get

$$
\tilde{u}(x,t;r,0) = (0.1r + 0.9)\left[x + \left(\frac{3x^{3-2\gamma}}{\Gamma(4-2\gamma)} - \frac{2x^{2-\gamma}}{\Gamma(3-\gamma)} \right)t + \left\{ \begin{array}{l} \dfrac{2\Gamma(4-\gamma)x^{3-2\gamma}}{\Gamma(3-\gamma)\Gamma(4-2\gamma)} - \\[2ex] \left(\dfrac{3\Gamma(5-2\gamma)}{\Gamma(4-2\gamma)} + \dfrac{\Gamma(5-\gamma)}{\Gamma(3-\gamma)} \right) \dfrac{x^{4-3\gamma}}{\Gamma(5-3\gamma)} \\[2ex] + \dfrac{3}{2} \dfrac{\Gamma(6-2\gamma)x^{5-4\gamma}}{\Gamma(4-2\gamma)\Gamma(6-4\gamma)} \end{array} \right\} \dfrac{t^2}{2} \right]. \tag{13.23}
$$

and

$$
\tilde{u}(x,t;r,1) = (1.1 - 0.1r)\left[x + \left(\frac{3x^{3-2\gamma}}{\Gamma(4-2\gamma)} - \frac{2x^{2-\gamma}}{\Gamma(3-\gamma)} \right)t + \left\{ \begin{array}{l} \dfrac{2\Gamma(4-\gamma)x^{3-2\gamma}}{\Gamma(3-\gamma)\Gamma(4-2\gamma)} - \\[2ex] \left(\dfrac{3\Gamma(5-2\gamma)}{\Gamma(4-2\gamma)} + \dfrac{\Gamma(5-\gamma)}{\Gamma(3-\gamma)} \right) \dfrac{x^{4-3\gamma}}{\Gamma(5-3\gamma)} \\[2ex] + \dfrac{3}{2} \dfrac{\Gamma(6-2\gamma)x^{5-4\gamma}}{\Gamma(4-2\gamma)\Gamma(6-4\gamma)} \end{array} \right\} \dfrac{t^2}{2} \right]. \tag{13.24}
$$

We note that in the special case when $r = 1$, the results (crisp) obtained by the proposed method are exactly the same as the solutions obtained by the method of Yildirim (2010).

Setting $r = 1$ and $\gamma = 1$ in Eqs. (13.23) and (13.24), one may get the solution of the aforementioned problem as

$$
u(x,t;) = x\left[1 + t + \frac{t^2}{2} + \cdots \right]. \tag{13.25}
$$

The aforementioned solution can be written in closed form as

$$
u(x,t;) = xe^t.
$$

Case 2   Next, we consider $A(x) = x/6$ and $B(x) = x^2/12$ with triangular fuzzy initial condition, namely $\tilde{u}(x,0;r) = x^2 [0.1r + 0.9, \ 1.1 - 0.1r]$.

Again, by applying the proposed procedure discussed previously, one may get the solution

$$\tilde{u}_0(x,t;r,\beta) = x^2\{\beta(0.2 - 0.2r) + (0.1r + 0.9)\}, \tag{13.26}$$

$$\tilde{u}_1(x,t;r,\beta) = \{\beta(0.2 - 0.2r) + (0.1r + 0.9)\}\left(\frac{2x^{4-2\gamma}}{\Gamma(5-2\gamma)} - \frac{x^{3-\gamma}}{\Gamma(4-\gamma)}\right)\frac{t^\alpha}{\Gamma(\alpha+1)}, \tag{13.27}$$

$$\tilde{u}_2(x,t;r,\beta) = \{\beta(0.2 - 0.2r) + (0.1r + 0.9)\}\left[\begin{array}{c}\left(\begin{array}{c}\dfrac{\Gamma(5-\gamma)}{6\Gamma(4-\gamma)}\dfrac{x^{4-2\gamma}}{\Gamma(5-2\gamma)}-\\[2mm]\left(\dfrac{\Gamma(6-2\gamma)}{3\Gamma(5-2\gamma)}+\\[2mm]\dfrac{1}{12}\dfrac{\Gamma(6-\gamma)}{\Gamma(4-\gamma)}\right)\dfrac{x^{5-3\gamma}}{\Gamma(6-3\gamma)}\end{array}\right)\dfrac{t^{2\alpha}}{\Gamma(2\alpha+1)}\\[6mm]+\dfrac{1}{6}\dfrac{\Gamma(7-2\gamma)x^{6-4\gamma}}{\Gamma(5-2\gamma)\Gamma(7-4\gamma)}\end{array}\right] \tag{13.28}$$

and so on. Substituting Eqs. (13.26)–(13.28) in Eq. (13.12), one may get the solution of $\tilde{u}(x,t;r,\beta)$ as

$$\tilde{u}(x,t;r,\beta) = \{\beta(0.2 - 0.2r) + (0.1r + 0.9)\}$$
$$\times\left[\begin{array}{c}x^2+\left(\dfrac{2x^{4-2\gamma}}{\Gamma(5-2\gamma)}-\dfrac{x^{3-\gamma}}{\Gamma(4-\gamma)}\right)\dfrac{t^\alpha}{\Gamma(\alpha+1)}+\\[4mm]\left(\begin{array}{c}\dfrac{\Gamma(5-\gamma)}{6\Gamma(4-\gamma)}\dfrac{x^{4-2\gamma}}{\Gamma(5-2\gamma)}-\\[2mm]\left(\dfrac{\Gamma(6-2\gamma)}{3\Gamma(5-2\gamma)}+\dfrac{1}{12}\dfrac{\Gamma(6-\gamma)}{\Gamma(4-\gamma)}\right)\dfrac{x^{5-3\gamma}}{\Gamma(6-3\gamma)}\\[2mm]+\dfrac{1}{6}\dfrac{\Gamma(7-2\gamma)x^{6-4\gamma}}{\Gamma(5-2\gamma)\Gamma(7-4\gamma)}\end{array}\right)\dfrac{t^{2\alpha}}{\Gamma(2\alpha+1)}\end{array}\right]. \tag{13.29}$$

For $\beta = 0$ and 1 in $\tilde{u}(x,t;r,\beta)$, we get the lower and upper bounds of the fuzzy solutions, respectively, as

$$\tilde{u}(x,t;r,0) = (0.1r + 0.9)\left[\begin{array}{c}x^2+\left(\dfrac{2x^{4-2\gamma}}{\Gamma(5-2\gamma)}-\dfrac{x^{3-\gamma}}{\Gamma(4-\gamma)}\right)\dfrac{t^\alpha}{\Gamma(\alpha+1)}+\\[4mm]\left(\begin{array}{c}\dfrac{\Gamma(5-\gamma)}{6\Gamma(4-\gamma)}\dfrac{x^{4-2\gamma}}{\Gamma(5-2\gamma)}-\\[2mm]\left(\dfrac{\Gamma(6-2\gamma)}{3\Gamma(5-2\gamma)}+\dfrac{1}{12}\dfrac{\Gamma(6-\gamma)}{\Gamma(4-\gamma)}\right)\dfrac{x^{5-3\gamma}}{\Gamma(6-3\gamma)}\\[2mm]+\dfrac{1}{6}\dfrac{\Gamma(7-2\gamma)x^{6-4\gamma}}{\Gamma(5-2\gamma)\Gamma(7-4\gamma)}\end{array}\right)\dfrac{t^{2\alpha}}{\Gamma(2\alpha+1)}\end{array}\right]. \tag{13.30}$$

and

$$
\tilde{u}(x, t; r, 1) = (1.1 - 0.1r) \left[\begin{array}{l} x^2 + \left(\dfrac{2x^{4-2\gamma}}{\Gamma(5-2\gamma)} - \dfrac{x^{3-\gamma}}{\Gamma(4-\gamma)} \right) \dfrac{t^\alpha}{\Gamma(\alpha+1)} + \\[2ex] \left(\dfrac{\Gamma(5-\gamma)}{6\Gamma(4-\gamma)} \dfrac{x^{4-2\gamma}}{\Gamma(5-2\gamma)} - \right. \\[2ex] \left. \left(\dfrac{\Gamma(6-2\gamma)}{3\Gamma(5-2\gamma)} + \dfrac{1}{12} \dfrac{\Gamma(6-\gamma)}{\Gamma(4-\gamma)} \right) \dfrac{x^{5-3\gamma}}{\Gamma(6-3\gamma)} \right. \dfrac{t^{2\alpha}}{\Gamma(2\alpha+1)} \\[2ex] \left. + \dfrac{1}{6} \dfrac{\Gamma(7-2\gamma)x^{6-4\gamma}}{\Gamma(5-2\gamma)\Gamma(7-4\gamma)} \right) \end{array} \right].
$$

(13.31)

The solution obtained by the proposed method for $r = 1$ (crisp result) is again found to be exactly the same as that of Yildirim (2010).

13.3.2 Using ADM

Uncertain solution bounds for Eq. (13.5) using ADM with double parametric form of fuzzy numbers have been discussed here. Depending upon $A(x)$ and $B(x)$, K, $F(x)$, two different Cases 3 and 4 have been analyzed in the following paragraphs.

Case 3 **(Same as Case 1 of HPM)** Here $\alpha = 1$, $A(x) = x$, and $B(x) = x^2/2$ with fuzzy initial condition in parametric form as $\tilde{u}(x, 0; r) = [0.1r + 0.9, 1.1 - 0.1r]x$ have been considered. Hence, Eq. (13.1) will become

$$
\frac{\partial \tilde{u}}{\partial t} = \left[-\frac{\partial^\gamma}{\partial x^\gamma} x + \frac{\partial^{2\gamma}}{\partial x^{2\gamma}} \frac{x^2}{2} \right] \tilde{u}(x, t) \quad t > 0, \quad x > 0, \quad 0 < \gamma \le 1,
$$

(13.32)

with fuzzy initial condition

$$
\tilde{u}(x, 0; r) = [0.1r + 0.9, 1.1 - 0.1r]x.
$$

Using double parametric form, Eq. (13.32) and the fuzzy initial condition become

$$
\frac{\partial^\alpha \tilde{u}(x, t; r, \beta)}{\partial t^\alpha} = \left[-\frac{\partial^\gamma}{\partial x^\gamma} x + \frac{\partial^{2\gamma}}{\partial x^{2\gamma}} \frac{x^2}{2} \right] \tilde{u}(x, t; r, \beta),
$$

with the initial condition

$$
\tilde{u}(x, 0; r, \beta) = \beta(0.2 - 0.2r) + (0.1r + 0.9).
$$

By using ADM, we have

$$
\tilde{u}_0(x, t; r, \beta) = \{\beta(0.2 - 0.2r) + (0.1r + 0.9)\}x,
$$

(13.33)

$$\tilde{u}_1(x, t; r, \beta) = \{\beta(0.2 - 0.2r) + (0.1r + 0.9)\}t \left[\frac{3x^{3-2\gamma}}{\Gamma(4 - 2\gamma)} - \frac{2x^{2-\gamma}}{\Gamma(3 - \gamma)} \right], \quad (13.34)$$

$$\tilde{u}_2(x, t; r, \beta) = \{\beta(0.2 - 0.2r) + (0.1r + 0.9)\}$$

$$\times \frac{t^2}{2} \left[\begin{array}{l} \dfrac{2\Gamma(4 - \gamma)}{\Gamma(3 - \gamma)\Gamma(4 - 2\gamma)} x^{3-2\gamma} - \\[2mm] \left(\dfrac{3\Gamma(5 - 2\gamma)}{\Gamma(4 - 2\gamma)} + \dfrac{\Gamma(5 - \gamma)}{\Gamma(3 - \gamma)} \right) \dfrac{x^{4-3\gamma}}{\Gamma(5 - 3\gamma)} + \\[2mm] \dfrac{3}{2} \dfrac{\Gamma(6 - 2\gamma) x^{5-4\gamma}}{\Gamma(4 - 2\gamma)\Gamma(6 - 4\gamma)} \end{array} \right], \quad (13.35)$$

and so on.

Therefore, the solution can be written as

$$\tilde{u}(x, t; r, \beta) = \{\beta(0.2 - 0.2r) + (0.1r + 0.9)\}$$

$$\times \left[\begin{array}{l} x + \left(\dfrac{3x^{3-2\gamma}}{\Gamma(4 - 2\gamma)} - \dfrac{2x^{2-\gamma}}{\Gamma(3 - \gamma)} \right) t + \\[3mm] \left\{ \begin{array}{l} \dfrac{2\Gamma(4 - \gamma) x^{3-2\gamma}}{\Gamma(3 - \gamma)\Gamma(4 - 2\gamma)} - \\[2mm] \left(\dfrac{3\Gamma(5 - 2\gamma)}{\Gamma(4 - 2\gamma)} + \dfrac{\Gamma(5 - \gamma)}{\Gamma(3 - \gamma)} \right) \dfrac{x^{4-3\gamma}}{\Gamma(5 - 3\gamma)} \\[2mm] + \dfrac{3}{2} \dfrac{\Gamma(6 - 2\gamma) x^{5-4\gamma}}{\Gamma(4 - 2\gamma)\Gamma(6 - 4\gamma)} \end{array} \right\} \dfrac{t^2}{2} \end{array} \right]. \quad (13.36)$$

To obtain the solution bound in single parametric form, we may put $\beta = 0$ and 1 to get the lower and upper bounds of the solution, respectively, as

$$\tilde{u}(x, t; r, 0) = (0.1r + 0.9) \left[\begin{array}{l} x + \left(\dfrac{3x^{3-2\gamma}}{\Gamma(4 - 2\gamma)} - \dfrac{2x^{2-\gamma}}{\Gamma(3 - \gamma)} \right) t + \\[3mm] \left\{ \begin{array}{l} \dfrac{2\Gamma(4 - \gamma) x^{3-2\gamma}}{\Gamma(3 - \gamma)\Gamma(4 - 2\gamma)} - \\[2mm] \left(\dfrac{3\Gamma(5 - 2\gamma)}{\Gamma(4 - 2\gamma)} + \dfrac{\Gamma(5 - \gamma)}{\Gamma(3 - \gamma)} \right) \dfrac{x^{4-3\gamma}}{\Gamma(5 - 3\gamma)} \\[2mm] + \dfrac{3}{2} \dfrac{\Gamma(6 - 2\gamma) x^{5-4\gamma}}{\Gamma(4 - 2\gamma)\Gamma(6 - 4\gamma)} \end{array} \right\} \dfrac{t^2}{2} \end{array} \right]. \quad (13.37)$$

and

$$\tilde{u}(x,t;r,1) = (1.1 - 0.1r) \begin{bmatrix} x + \left(\dfrac{3x^{3-2\gamma}}{\Gamma(4-2\gamma)} - \dfrac{2x^{2-\gamma}}{\Gamma(3-\gamma)} \right) t + \\ \left\{ \begin{aligned} & \dfrac{2\Gamma(4-\gamma)x^{3-2\gamma}}{\Gamma(3-\gamma)\Gamma(4-2\gamma)} - \\ & \left(\dfrac{3\Gamma(5-2\gamma)}{\Gamma(4-2\gamma)} + \dfrac{\Gamma(5-\gamma)}{\Gamma(3-\gamma)} \right) \dfrac{x^{4-3\gamma}}{\Gamma(5-3\gamma)} \\ & + \dfrac{3}{2}\dfrac{\Gamma(6-2\gamma)x^{5-4\gamma}}{\Gamma(4-2\gamma)\Gamma(6-4\gamma)} \end{aligned} \right\} \dfrac{t^2}{2} \end{bmatrix} . \quad (13.38)$$

One may note that in the special case when $r = 1$, the results (crisp) obtained by the proposed method are exactly the same as the solution obtained by the method of Yildirim (2010).

Case 4 (Same as Case B of HPM) Again, by applying the procedure discussed previously, we get the solution

$$\tilde{u}_0(x,t;r,\beta) = x^2\{\beta(0.2 - 0.2r) + (0.1r + 0.9)\}, \quad (13.39)$$

$$\tilde{u}_1(x,t;r,\beta) = \{\beta(0.2 - 0.2r) + (0.1r + 0.9)\} \left(\frac{2x^{4-2\gamma}}{\Gamma(5-2\gamma)} - \frac{x^{3-\gamma}}{\Gamma(4-\gamma)} \right) \frac{t^\alpha}{\Gamma(\alpha+1)}, \quad (13.40)$$

$$\tilde{u}_2(x,t;r,\beta) = \{\beta(0.2 - 0.2r) + (0.1r + 0.9)\}$$

$$\times \begin{bmatrix} \dfrac{\Gamma(5-\gamma)}{6\Gamma(4-\gamma)}\dfrac{x^{4-2\gamma}}{\Gamma(5-2\gamma)} - \\ \left(\begin{aligned} & \dfrac{\Gamma(6-2\gamma)}{3\Gamma(5-2\gamma)} + \\ & \dfrac{1}{12}\dfrac{\Gamma(6-\gamma)}{\Gamma(4-\gamma)} \end{aligned} \right) \dfrac{x^{5-3\gamma}}{\Gamma(6-3\gamma)} \dfrac{t^{2\alpha}}{\Gamma(2\alpha+1)} \\ + \dfrac{1}{6}\dfrac{\Gamma(7-2\gamma)x^{6-4\gamma}}{\Gamma(5-2\gamma)\Gamma(7-4\gamma)} \end{bmatrix}, \quad (13.41)$$

and so on. Substituting Eqs. (13.39)–(13.41) in Eq. (13.17), one may get the solution of $\tilde{u}(x,t;r,\beta)$ as

$$\widetilde{u}(x,t;r,\beta) = \{\beta(0.2 - 0.2r) + (0.1r + 0.9)\}$$

$$\times \left[x^2 + \left(\frac{2x^{4-2\gamma}}{\Gamma(5-2\gamma)} - \frac{x^{3-\gamma}}{\Gamma(4-\gamma)} \right) \frac{t^\alpha}{\Gamma(\alpha+1)} + \left(\begin{array}{l} \frac{\Gamma(5-\gamma)}{6\Gamma(4-\gamma)} \frac{x^{4-2\gamma}}{\Gamma(5-2\gamma)} - \\ \left(\frac{\Gamma(6-2\gamma)}{3\Gamma(5-2\gamma)} + \frac{1}{12}\frac{\Gamma(6-\gamma)}{\Gamma(4-\gamma)} \right) \frac{x^{5-3\gamma}}{\Gamma(6-3\gamma)} \\ + \frac{1}{6}\frac{\Gamma(7-2\gamma)x^{6-4\gamma}}{\Gamma(5-2\gamma)\Gamma(7-4\gamma)} \end{array} \right) \frac{t^{2\alpha}}{\Gamma(2\alpha+1)} \right]. \quad (13.42)$$

Putting $\beta = 0$ and 1 in $\widetilde{u}(x,t;r,\beta)$, the lower and upper bounds of the fuzzy solutions have been obtained, respectively, as

$$\widetilde{u}(x,t;r,0) = (0.1r + 0.9) \left[x^2 + \left(\frac{2x^{4-2\gamma}}{\Gamma(5-2\gamma)} - \frac{x^{3-\gamma}}{\Gamma(4-\gamma)} \right) \frac{t^\alpha}{\Gamma(\alpha+1)} + \left(\begin{array}{l} \frac{\Gamma(5-\gamma)}{6\Gamma(4-\gamma)} \frac{x^{4-2\gamma}}{\Gamma(5-2\gamma)} - \\ \left(\frac{\Gamma(6-2\gamma)}{3\Gamma(5-2\gamma)} + \frac{1}{12}\frac{\Gamma(6-\gamma)}{\Gamma(4-\gamma)} \right) \frac{x^{5-3\gamma}}{\Gamma(6-3\gamma)} \\ + \frac{1}{6}\frac{\Gamma(7-2\gamma)x^{6-4\gamma}}{\Gamma(5-2\gamma)\Gamma(7-4\gamma)} \end{array} \right) \frac{t^{2\alpha}}{\Gamma(2\alpha+1)} \right].$$

$$(13.43)$$

and

$$\widetilde{u}(x,t;r,1) = (1.1 - 0.1r) \left[x^2 + \left(\frac{2x^{4-2\gamma}}{\Gamma(5-2\gamma)} - \frac{x^{3-\gamma}}{\Gamma(4-\gamma)} \right) \frac{t^\alpha}{\Gamma(\alpha+1)} + \left(\begin{array}{l} \frac{\Gamma(5-\gamma)}{6\Gamma(4-\gamma)} \frac{x^{4-2\gamma}}{\Gamma(5-2\gamma)} - \\ \left(\frac{\Gamma(6-2\gamma)}{3\Gamma(5-2\gamma)} + \frac{1}{12}\frac{\Gamma(6-\gamma)}{\Gamma(4-\gamma)} \right) \frac{x^{5-3\gamma}}{\Gamma(6-3\gamma)} \\ + \frac{1}{6}\frac{\Gamma(7-2\gamma)x^{6-4\gamma}}{\Gamma(5-2\gamma)\Gamma(7-4\gamma)} \end{array} \right) \frac{t^{2\alpha}}{\Gamma(2\alpha+1)} \right].$$

$$(13.44)$$

The solution obtained by the proposed method for $r = 1$ is again found to be exactly the same as that of (crisp result) Yildirim (2010).

13.4 NUMERICAL RESULTS OF FUZZY FRACTIONAL FOKKER–PLANCK EQUATION

Numerical results for fuzzy fractional Fokker–Planck equation with different $A(x)$, $B(x)$ and fuzzy initial conditions are computed by HPM and ADM. Also, it is

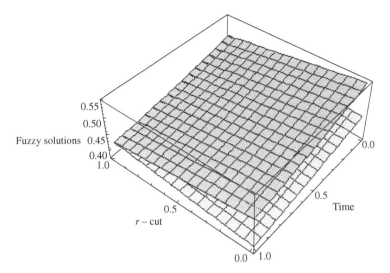

Figure 13.1   Fuzzy solution for Case 1 using HPM

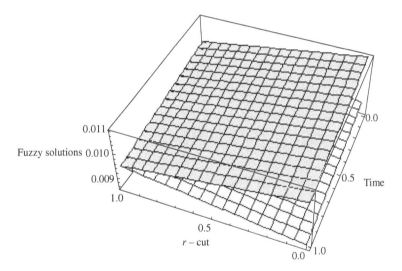

Figure 13.2   Fuzzy solution for Case 2 using HPM

interesting to note that the results by both the methods are same. The results obtained by the present analysis are compared with the existing solution of Yildirim (2010) in special cases to show the validation of the proposed analysis.

Triangular fuzzy solutions for Cases 1 and 2 are depicted in Figs. 13.1 and 13.2, respectively, by varying the time from 0 to 1 for $\gamma = 0.5$ and $\alpha = 1$ using HPM. It may be worth mentioning that for all the cases, the present results exactly agree with the solution of Yildirim (2010) in the special case of $r = 1$.

TABLE 13.1 Comparison of HPM, ADM, and Yildirim (2010) Solutions of Cases 1 and 3 for $x = 0.5$, $\gamma = 0.5$, and $\alpha = 1$

t	$\underline{u}(x,t)$ $r = 0.6$		$\overline{u}(x,t)$ $r = 0.6$		$\underline{u}(x,t) = \overline{u}(x,t) = u(x,t)$ $r = 1$		$u(x,t)$
	HPM	ADM	HPM	ADM	HPM	ADM	Yildirim (2010)
0	0.48	0.48	0.52	0.52	0.5	0.5	0.5
0.2	0.45094	0.45094	0.488521	0.488521	0.46973	0.46973	0.46973
0.4	0.42402	0.42402	0.45936	0.45936	0.44169	0.44169	0.44169
0.6	0.39926	0.39926	0.43253	0.43253	0.41589	0.41589	0.41589
0.8	0.37663	0.37663	0.408022	0.408022	0.39232	0.39232	0.39232
1	0.35615	0.35615	0.38583	0.38583	0.37099	0.37099	0.37099

TABLE 13.2 Comparison of HPM, ADM, and Yildirim (2010) Solutions of Cases 2 and 4 for $x = 0.5$, $\gamma = 0.5$, and $\alpha = 1$

t	$\underline{u}(x,t)$ $r = 0.6$		$\overline{u}(x,t)$ $r = 0.6$		$\underline{u}(x,t) = \overline{u}(x,t) = u(x,t)$ $r = 1$		$u(x,t)$
	HPM	ADM	HPM	ADM	HPM	ADM	Yildirim (2010)
0	0.24	0.24	0.26	0.26	0.25	0.25	0.25
0.2	0.23780	0.23780	0.25761	0.25761	0.24770	0.24770	0.24770
0.4	0.23563	0.23563	0.25526	0.25526	0.24544	0.24544	0.24544
0.6	0.23348	0.23348	0.25294	0.25294	0.24321	0.24321	0.24321
0.8	0.23137	0.23137	0.25065	0.25065	0.24101	0.24101	0.24101
1	0.22928	0.22928	0.24839	0.24839	0.23883	0.23883	0.23883

Computed results in digital form for the Cases 1 and 3 are tabulated in Table 13.1. Similarly obtained results for Cases 2 and 4 are given in Table 13.2. In these cases, t varies from 0 to 1. Here we consider $x = 0.5$, $\gamma = 0.6$, and $\alpha = 1$. It is again interesting to note that in the special case (for $r = 1$), the obtained results by different methods exactly match with that of Yildirim (2010).

In this chapter, HPM and ADM have successfully been applied to obtain the solution of fuzzy fractional Fokker–Planck equation with initial conditions as triangular fuzzy number using double parametric form of fuzzy numbers. The proposed double parametric form approach is found to be easy and straightforward, similarly to the previous chapters.

BIBLIOGRAPHY

Abbaoui K, Cherruault Y. Convergence of Adomian's method applied to different equations. *Comput Math Appl* 1994;**28**:103–109.

Abbaoui K, Cherruault Y. New ideas for proving convergence of decomposition methods. *Comput Math Appl* 1995;**29**:103–108.

Adomian G. A new approach to nonlinear partial differential equations. *J Math Anal Appl* 1984;**102**:420–434.

Adomian G. *Solving Frontier Problems of Physics: The Decomposition Method.* Boston: Kluwer Academic Publishers; 1994.

Cetinkaya A, Kiymaz O. The solution of the time-fractional diffusion equation by the generalized differential transform method. *Math Comput Model* 2013;**57**:2349–2354.

Cherruault Y. Convergence of Adomian's method. *Kybernetes* 1989;**18**:31–38.

Chen YM, Wu YB. Wavelet method for a class of fractional convection–diffusion equation with variable coefficients. *J Comput Sci* 2010;**1**:146–149.

Frank TD. Stochastic feedback, nonlinear families of Markov processes, and nonlinear Fokker–Planck equations. *Physica* 2004;**A331**:391–408.

Garg M, Manohar P. Numerical solution of fractional diffusion-wave equation with two space variables by matrix method. *Fract Calc Appl Anal* 2010;**13**:191–207.

Godal MA, Salah A, Khan M, Batool SI. A novel analytical solution of a fractional diffusion problem by homotopy analysis transform method. *Neural Comput Appl* 2013;**23**:1643–1647.

Gorenflo R, Luchko Y, Mainardi F. Wright functions as scale-invariant solutions of the diffusion-wave equation. *J Comput Appl Math* 2000;**118**:175–191.

He JH. Homotopy perturbation technique. *Comput Methods Appl Mech Eng* 1999; **178**:257–262.

He JH. A coupling method of homotopy technique and perturbation technique for nonlinear problems. *Int J Nonlinear Mech* 2000;**35**:37–43.

Himoun N, Abbaoui K, Cherruault Y. New results of convergence of Adomian's method. *Kyber* 1999;**28**:423–429.

Yildirim A. Analytical approach to Fokker–Planck equation with space- and time-fractional derivatives by means of the homotopy perturbation method. *J King Saud Univ (Sci)* 2010;**22**:257–264.

Tapaswini S, Chakraverty S. Non-probabilistic solutions of uncertain fractional order diffusion equations. *Fund Inform* 2014;**133**:19–34.

14

FUZZY FRACTIONAL BAGLEY–TORVIK EQUATIONS

The target of this chapter is to investigate the solution of fuzzy fractional Bagley–Torvik equations using homotopy perturbation method (HPM). Involved initial conditions are considered as uncertain and are taken in terms of triangular and trapezoidal fuzzy numbers. In order to show the applicability of the proposed method, we have considered a model of motion of a rigid plate immersed in a Newtonian fluid.

14.1 VARIOUS TYPES OF FUZZY FRACTIONAL BAGLEY–TORVIK EQUATIONS

In this section, HPM has been applied to solve fuzzy fractional Bagley–Torvik equations. Various types of fuzzy fractional Bagley–Torvik equations are addressed as Cases 1–3 and are explained as follows.

Case 1 Let us consider an inhomogeneous fuzzy fractional Bagley–Torvik equation (EL-Sayed et al., 2004) as

$$A\widetilde{y}''(t) + BD^{3/2}\widetilde{y}(t) + C\widetilde{y}(t) = 8, \quad 0 \le t \le 1 \tag{14.1}$$

with initial condition in terms of triangular fuzzy number, namely $\widetilde{y}(0) = (-0.1, \ 0, \ 0.1)$.

Fuzzy Arbitrary Order System: Fuzzy Fractional Differential Equations and Applications, First Edition.
Snehashish Chakraverty, Smita Tapaswini, and Diptiranjan Behera.
© 2016 John Wiley & Sons, Inc. Published 2016 by John Wiley & Sons, Inc.

Using r-cut approach, the triangular fuzzy initial condition becomes

$$\widetilde{y}(0) = [0.1r - 0.1, 0.1 - 0.1r], \quad 0 \le r \le 1.$$

Here $\widetilde{y}$ is a fuzzy function of t. It may be noted that fuzzy fractional Bagley–Torvik equation may be reduced to a set of ordinary fractional differential equations by Hukuhara derivative (Bede, 2008),

$$A\underline{y}''(t) + BD^{3/2}\underline{y}(t) + C\underline{y}(t) = 8, \quad 0 \le t \le 1, \tag{14.2}$$

$$A\overline{y}''(t) + BD^{3/2}\overline{y}(t) + C\overline{y}(t) = 8, \quad 0 \le t \le 1. \tag{14.3}$$

As mentioned earlier, in homotopy technique, we construct a homotopy (He, 1999, 2000) for Eqs. (14.2) and (14.3) as follows:

$$H\left(\underline{Y}, p\right) = (1 - p)\left[\frac{d^2\underline{Y}}{dt^2} - \frac{d^2\underline{y}_0}{d^2t}\right] + p\left[\frac{d^2\underline{Y}}{dt^2} + \frac{B}{A}\frac{d^{3/2}\underline{Y}(t)}{dt^{3/2}} + \frac{C}{A}\underline{Y}(t) - \frac{8}{A}\right] = 0, \tag{14.4}$$

$$H\left(\overline{Y}, p\right) = (1 - p)\left[\frac{d^2\overline{Y}}{dt^2} - \frac{d^2\overline{y}_0}{d^2t}\right] + p\left[\frac{d^2\overline{Y}}{dt^2} + \frac{B}{A}\frac{d^{3/2}\overline{Y}(t)}{dt^{3/2}} + \frac{C}{A}\overline{Y}(t) - \frac{8}{A}\right] = 0. \tag{14.5}$$

Let us assume the solutions of Eqs. (14.4) and (14.5) as a power series in p, respectively, as

$$\underline{Y}(t; r) = \underline{Y}_0(t; r) + p\underline{Y}_1(t; r) + p^2\underline{Y}_2(t; r) + \cdots \tag{14.6}$$

and

$$\overline{Y}(t; r) = \overline{Y}_0(t; r) + p\overline{Y}_1(t; r) + p^2\overline{Y}_2(t; r) + \cdots, \tag{14.7}$$

where, $\underline{Y}_i(t; r)$ and $\overline{Y}_i(t; r)$ for $i = 0, 1, 2, \ldots$ are functions yet to be determined. Substituting Eqs. (14.6) and (14.7) into Eqs. (14.4) and (14.5), respectively, and equating the terms with the identical powers of p, we have

$$p^0 : \begin{cases} \dfrac{d^2\underline{Y}_0}{dt^2} - \dfrac{d^2\underline{y}_0}{dt^2} = 0, \\[2ex] \dfrac{d^2\overline{Y}_0}{dt^2} - \dfrac{d^2\overline{y}_0}{dt^2} = 0, \end{cases}$$

$$p^1 : \begin{cases} \dfrac{d^2\underline{Y}_1}{dt^2} + \dfrac{d^2\underline{y}_0}{dt^2} + \dfrac{B}{A}\dfrac{d^{3/2}\underline{Y}_0(t)}{dt^{3/2}} + \dfrac{C}{A}\underline{Y}_0 - \dfrac{8}{A} = 0, \\[2ex] \dfrac{d^2\overline{Y}_1}{dt^2} + \dfrac{d^2\overline{y}_0}{dt^2} + \dfrac{B}{A}\dfrac{d^{3/2}\overline{Y}_0}{dt^{3/2}} + \dfrac{C}{A}\overline{Y}_0 - \dfrac{8}{A} = 0, \end{cases}$$

$$p^2 : \begin{cases} \dfrac{d^2 \underline{Y}_2}{dt^2} + \dfrac{B}{A}\dfrac{d^{3/2}\underline{Y}_1(t)}{dt^{3/2}} + \dfrac{C}{A}\underline{Y}_1 = 0, \\[4mm] \dfrac{d^2 \overline{Y}_2}{dt^2} + \dfrac{B}{A}\dfrac{d^{3/2}\overline{Y}_1}{dt^{3/2}} + \dfrac{C}{A}\overline{Y}_1 = 0, \end{cases}$$

$$p^3 : \begin{cases} \dfrac{d^2 \underline{Y}_3}{dt^2} + \dfrac{B}{A}\dfrac{d^{3/2}\underline{Y}_2}{dt^{3/2}} + \dfrac{C}{A}\underline{Y}_2 = 0, \\[4mm] \dfrac{d^2 \overline{Y}_3}{dt^2} + \dfrac{B}{A}\dfrac{d^{3/2}\overline{Y}_2}{dt^{3/2}} + \dfrac{C}{A}\overline{Y}_2 = 0, \end{cases}$$

and so on.

The initial approximation $\underline{Y}_0(t)$ or $\underline{y}_0(t)$ and $\overline{Y}_0(t)$ or $\overline{y}_0(t)$ can be freely chosen. Here we set

$$\underline{Y}_0(t;r) = \underline{y}_0(t;r) = 0.1r - 0.1 \text{ and } \overline{Y}_0(t) = \overline{y}_0(t) = 0.1 - 0.1r.$$

Using the assumed initial condition into p^i terms for $i = 0, 1, 2, \ldots$, respectively, yields

$$\underline{Y}_1(t;r) = -\frac{C}{A}(0.1r - 0.1)\frac{t^2}{\Gamma(3)} + \frac{8t^2}{\Gamma(3)},$$

$$\overline{Y}_1(t;r) = -\frac{C}{A}(0.1 - 0.1r)\frac{t^2}{\Gamma(3)} + \frac{8t^2}{\Gamma(3)},$$

$$\underline{Y}_2(t;r) = \frac{BC}{A^2}(0.1r - 0.1)\frac{t^{5/2}}{\Gamma(7/2)} - \frac{8Bt^{5/2}}{A^2\Gamma(7/2)} + (0.1r - 0.1)\left(\frac{C}{A}\right)^2\frac{t^4}{\Gamma(5)}$$
$$- \frac{8C}{A^2}\frac{t^4}{\Gamma(5)},$$

$$\overline{Y}_2(t;r) = \frac{BC}{A^2}(0.1 - 0.1r)\frac{t^{5/2}}{\Gamma(7/2)} - \frac{8Bt^{5/2}}{A^2\Gamma(7/2)} + (0.1r - 0.1)\left(\frac{C}{A}\right)^2\frac{t^4}{\Gamma(5)}$$
$$- \frac{8C}{A^2}\frac{t^4}{\Gamma(5)},$$

$$\underline{Y}_3(t;r) = -\frac{8B^2C}{A^3}(0.1r - 0.1)\frac{t^3}{\Gamma(4)} + \frac{8B^2t^3}{A^3\Gamma(4)} - \frac{2BC^2}{A^3}(0.1r - 0.1)\frac{t^{9/2}}{\Gamma(11/2)}$$
$$+ \frac{16BCt^{9/2}}{A^3\Gamma(11/2)} + (0.1r - 0.1)\left(\frac{C}{A}\right)^3\frac{t^6}{\Gamma(7)} + \frac{8C^2}{A^3}\frac{t^6}{\Gamma(7)},$$

$$\overline{Y}_3(t;r) = -\frac{8B^2C}{A^3}(0.1-0.1r)\frac{t^3}{\Gamma(4)} + \frac{8B^2t^3}{A^3\Gamma(4)} - \frac{2BC^2}{A^3}(0.1-0.1r)\frac{t^{9/2}}{\Gamma(11/2)}$$

$$+ \frac{16BCt^{9/2}}{A^3\Gamma(11/2)} + (0.1-0.1r)\left(\frac{C}{A}\right)^3\frac{t^6}{\Gamma(7)} + \frac{8C^2}{A^3}\frac{t^6}{\Gamma(7)},$$

and so on. The fourth-order approximate solutions of Eqs. (14.2) and (14.3) may be written as

$$\underline{y}(t;r) \cong \underline{Y}_0(t;r) + \underline{Y}_1(t;r) + \underline{Y}_2(t;r) + \underline{Y}_3(t;r)$$

$$= -\frac{C}{A}(0.1r-0.1)\frac{t^2}{\Gamma(3)} + \frac{8t^2}{\Gamma(3)} + \frac{BC}{A^2}(0.1r-0.1)\frac{t^{5/2}}{\Gamma(7/2)} - \frac{8Bt^{5/2}}{A^2\Gamma(7/2)}$$

$$+ (0.1r-0.1)\left(\frac{C}{A}\right)^2\frac{t^4}{\Gamma(5)} - \frac{8C}{A^2}\frac{t^4}{\Gamma(5)} - \frac{8B^2C}{A^3}(0.1r-0.1)\frac{t^3}{\Gamma(4)} + \frac{8B^2}{A^3}\frac{t^3}{\Gamma(4)}$$

$$- \frac{2BC^2}{A^3}(0.1r-0.1)\frac{t^{9/2}}{\Gamma(11/2)} + \frac{16BC}{A^3}\frac{t^{9/2}}{\Gamma(11/2)} + (0.1r-0.1)\left(\frac{C}{A}\right)^3\frac{t^6}{\Gamma(7)}$$

$$+ \frac{8C^2}{A^3}\frac{t^6}{\Gamma(7)}$$

and

$$\overline{y}(t;r) \cong \overline{Y}_0(t;r) + \overline{Y}_1(t;r) + \overline{Y}_2(t;r) + \overline{Y}_3(t;r)$$

$$= -\frac{C}{A}(0.1-0.1r)\frac{t^2}{\Gamma(3)} + \frac{8t^2}{\Gamma(3)} + \frac{BC}{A^2}(0.1-0.1r)\frac{t^{5/2}}{\Gamma(7/2)} - \frac{8Bt^{5/2}}{A^2\Gamma(7/2)}$$

$$+ (0.1r-0.1)\left(\frac{C}{A}\right)^2\frac{t^4}{\Gamma(5)} - \frac{8C}{A^2}\frac{t^4}{\Gamma(5)} - \frac{8B^2C}{A^3}(0.1-0.1r)\frac{t^3}{\Gamma(4)} + \frac{8B^2}{A^3}\frac{t^3}{\Gamma(4)}$$

$$- \frac{2BC^2}{A^3}(0.1-0.1r)\frac{t^{9/2}}{\Gamma(11/2)} + \frac{16BC}{A^3}\frac{t^{9/2}}{\Gamma(11/2)} + (0.1-0.1r)\left(\frac{C}{A}\right)^3\frac{t^6}{\Gamma(7)}$$

$$+ \frac{8C^2}{A^3}\frac{t^6}{\Gamma(7)}.$$

Case 2   Next, we consider another type of fuzzy fractional Bagley–Torvik equation (EL-Sayed et al., 2004)

$$\widetilde{y}''(t) + D^{3/2}\widetilde{y}(t) + \widetilde{y}(t) = 1 + t, \quad 0 \le t \le 1 \tag{14.8}$$

with initial condition in terms of triangular fuzzy number, namely $\widetilde{y}(0) = (-0.1, 0, 0.1)$.

Again, using r-cut approach, the triangular fuzzy initial condition becomes

$$\widetilde{y}(0; r) = [0.1r - 0.1, 0.1 - 0.1r], \quad 0 \le r \le 1.$$

The aforementioned fuzzy fractional differential equation may similarly be reduced (as in Case 1) to a set of ordinary differential equations as follows:

$$\underline{y}''(t) + D^{3/2}\underline{y}(t) + \underline{y}(t) = 1 + t, \quad 0 \le t \le 1, \tag{14.9}$$

$$\overline{y}''(t) + D^{3/2}\overline{y}(t) + \overline{y}(t) = 1 + t, \quad 0 \le t \le 1. \tag{14.10}$$

By the homotopy technique, we construct a homotopy for Eqs. (14.9) and (14.10) as follows:

$$H(\underline{Y}, p) = (1 - p)\left[\frac{d^2\underline{Y}}{d^2t} - \frac{d^2\underline{y}_0}{d^2t}\right] + p\left[\frac{d^2\underline{Y}}{d^2t} + \frac{d^{3/2}\underline{Y}(t)}{dt^{3/2}} + \underline{Y}(t) - (1 + t)\right] = 0, \tag{14.11}$$

$$H(\overline{Y}, p) = (1 - p)\left[\frac{d^2\overline{Y}}{d^2t} - \frac{d^2\overline{y}}{d^2t}\right] + p\left[\frac{d^2\overline{Y}}{d^2t} + \frac{d^{3/2}\overline{Y}(t)}{dt^{3/2}} + \overline{Y}(t) - (1 + t)\right] = 0. \tag{14.12}$$

One may try to obtain a solution of Eqs. (14.11) and (14.12) in the form

$$\underline{Y}(t; r) = \underline{Y}_0(t; r) + p\underline{Y}_1(t; r) + p^2\underline{Y}_2(t; r) + \cdots, \tag{14.13}$$

$$\overline{Y}(t; r) = \overline{Y}_0(t; r) + p\overline{Y}_1(t; r) + p^2\overline{Y}_2(t; r) + \cdots \tag{14.14}$$

where, $\underline{Y}_i(t)$ and $\overline{Y}_i(t)$ for $i = 0, 1, 2, \ldots$ are functions yet to be determined. Substituting Eqs. (14.13) and (14.14) into Eqs. (14.11) and (14.12), respectively, and equating the terms with the identical powers of p, we have

$$p^0 : \begin{cases} \dfrac{d^2\underline{Y}_0}{dt^2} - \dfrac{d^2\underline{y}_0}{dt^2} = 0, \\[3mm] \dfrac{d^2\overline{Y}_0}{dt^2} - \dfrac{d^2\overline{y}_0}{dt^2} = 0, \end{cases}$$

$$p^1 : \begin{cases} \dfrac{d^2\underline{Y}_1}{dt^2} + \dfrac{d^2\underline{y}_0}{dt^2} + \dfrac{d^{3/2}\underline{Y}_0}{dt^{3/2}} + \underline{Y}_0 - (1 + t) = 0, \\[3mm] \dfrac{d^2\overline{Y}_1}{dt^2} + \dfrac{d^2\overline{y}_0}{dt^2} + \dfrac{d^{3/2}\overline{Y}_0}{dt^{3/2}} + \overline{Y}_0 - (1 + t) = 0, \end{cases}$$

$$p^2 : \begin{cases} \dfrac{d^2 \underline{Y}_2}{dt^2} + \dfrac{d^{3/2} \underline{Y}_1}{dt^{3/2}} + \underline{Y}_1 = 0, \\[3mm] \dfrac{d^2 \overline{Y}_2}{dt^2} + \dfrac{d^{3/2} \overline{Y}_1}{dt^{3/2}} + \overline{Y}_1 = 0, \end{cases}$$

$$p^3 : \begin{cases} \dfrac{d^2 \underline{Y}_3}{dt^2} + \dfrac{d^{3/2} \underline{Y}_2}{dt^{3/2}} + \underline{Y}_2 = 0, \\[3mm] \dfrac{d^2 \overline{Y}_3}{dt^2} + \dfrac{d^{3/2} \overline{Y}_2}{dt^{3/2}} + \overline{Y}_2 = 0, \end{cases}$$

and so on. The initial approximation $\underline{Y}_0(t)$ or $\underline{y}_0(t)$ and $\overline{Y}_0(t)$ or $\overline{y}_0(t)$ can be freely chosen. Here we set $\underline{Y}_0(t; r) = \underline{y}_0(t; r) = 0.1r - 0.1$ and $\overline{Y}_0(t; r) = \overline{y}_0(t; r) = 0.1 - 0.1r$.

Substituting the assumed initial condition into p^i terms, respectively, yields

$$\underline{Y}_1(t; r) = -(0.1r - 0.1)\frac{t^2}{\Gamma(3)} + \frac{t^2}{\Gamma(3)} + \frac{t^3}{\Gamma(4)},$$

$$\overline{Y}_1(t; r) = -(0.1 - 0.1r)\frac{t^2}{\Gamma(3)} + \frac{t^2}{\Gamma(3)} + \frac{t^3}{\Gamma(4)},$$

$$\underline{Y}_2(t; r) = (0.1r - 0.1)\frac{t^{5/2}}{\Gamma(7/2)} - \frac{t^{5/2}}{\Gamma(7/2)} - \frac{t^{7/2}}{\Gamma(9/2)} + (0.1r - 0.1)\frac{t^4}{\Gamma(5)}$$
$$- \frac{t^4}{\Gamma(5)} - \frac{t^5}{\Gamma(6)},$$

$$\overline{Y}_2(t; r) = (0.1 - 0.1r)\frac{t^{5/2}}{\Gamma(7/2)} - \frac{t^{5/2}}{\Gamma(7/2)} - \frac{t^{7/2}}{\Gamma(9/2)} + (0.1 - 0.1r)\frac{t^4}{\Gamma(5)}$$
$$- \frac{t^4}{\Gamma(5)} - \frac{t^5}{\Gamma(6)},$$

$$\underline{Y}_3(t; r) = -(0.1r - 0.1)\frac{t^3}{\Gamma(4)} + \frac{t^3}{\Gamma(4)} + \frac{t^4}{\Gamma(5)} - 2(0.1r - 0.1)\frac{t^{9/2}}{\Gamma(11/2)} + \frac{2t^{9/2}}{\Gamma(11/2)}$$
$$+ \frac{2t^{11/2}}{\Gamma(13/2)} - (0.1r - 0.1)\frac{t^6}{\Gamma(7)} + \frac{t^6}{\Gamma(7)} + \frac{t^7}{\Gamma(8)},$$

$$\overline{Y}_3(t; r) = -(0.1 - 0.1r)\frac{t^3}{\Gamma(4)} + \frac{t^3}{\Gamma(4)} + \frac{t^4}{\Gamma(5)} - 2(0.1 - 0.1r)\frac{t^{9/2}}{\Gamma(11/2)} + \frac{2t^{9/2}}{\Gamma(11/2)}$$
$$+ \frac{2t^{11/2}}{\Gamma(13/2)} - (0.1 - 0.1r)\frac{t^6}{\Gamma(7)} + \frac{t^6}{\Gamma(7)} + \frac{t^7}{\Gamma(8)},$$

and so on. Again, the fourth-order approximate solution of Eqs. (14.9) and (14.10) may be written in a finite series form as

$$\underline{y}(t;r) \cong \underline{Y}_0(t;r) + \underline{Y}_1(t;r) + \underline{Y}_2(t;r) + \underline{Y}_3(t;r)$$

$$= (0.1r - 0.1) - (0.1r - 0.1)\frac{t^2}{\Gamma(3)} + \frac{t^2}{\Gamma(3)} + (0.1r - 0.1)\frac{t^{5/2}}{\Gamma(7/2)} - \frac{t^{5/2}}{\Gamma(7/2)}$$

$$- \frac{t^{7/2}}{\Gamma(9/2)} + (0.1r - 0.1)\frac{t^4}{\Gamma(5)} - \frac{t^5}{\Gamma(6)} - (0.1r - 0.1)\frac{t^3}{\Gamma(4)} + \frac{t^3}{\Gamma(4)} + \frac{t^4}{\Gamma(5)}$$

$$- 2(0.1r - 0.1)\frac{t^{9/2}}{\Gamma(11/2)} + \frac{2t^{9/2}}{\Gamma(11/2)} + \frac{2t^{11/2}}{\Gamma(13/2)} - (0.1r - 0.1)\frac{t^6}{\Gamma(7)}$$

$$+ \frac{t^6}{\Gamma(7)} + \frac{t^7}{\Gamma(8)}$$

and

$$\overline{y}(t;r) \cong \overline{Y}_0(t;r) + \overline{Y}_1(t;r) + \overline{Y}_2(t;r) + \overline{Y}_3(t;r)$$

$$= (0.1 - 0.1r) - (0.1 - 0.1r)\frac{t^2}{\Gamma(3)} + \frac{t^2}{\Gamma(3)} + (0.1 - 0.1r)\frac{t^{5/2}}{\Gamma(7/2)} - \frac{t^{5/2}}{\Gamma(7/2)}$$

$$- \frac{t^{7/2}}{\Gamma(9/2)} + (0.1 - 0.1r)\frac{t^4}{\Gamma(5)} - \frac{t^5}{\Gamma(6)} - (0.1 - 0.1r)\frac{t^3}{\Gamma(4)} + \frac{t^3}{\Gamma(4)} + \frac{t^4}{\Gamma(5)}$$

$$- 2(0.1 - 0.1r)\frac{t^{9/2}}{\Gamma(11/2)} + \frac{2t^{9/2}}{\Gamma(11/2)} + \frac{2t^{11/2}}{\Gamma(13/2)} - (0.1 - 0.1r)\frac{t^6}{\Gamma(7)}$$

$$+ \frac{t^6}{\Gamma(7)} + \frac{t^7}{\Gamma(8)}.$$

Case 3 Finally, the inhomogeneous nonlinear fractional Bagley–Torvik equation (EL-Sayed et al., 2004) is considered as

$$A\widetilde{y}''(t) + BD^{3/2}\widetilde{y}(t) + C\widetilde{y}^3(t) = f(t), \quad t > 0, \quad \text{where } f(t) = 8 \text{ for } 0 \le t \le 1,$$
$$(14.15)$$

subject to

$$\widetilde{y}(0) = \widetilde{y}'(0) = (-0.1, 0, 0.1).$$

Using Hukuhara derivative (Bede, 2008), the nonlinear fuzzy fractional Bagley–Torvik differential equation may also be reduced to

$$A\underline{y}''(t) + BD^{3/2}\underline{y}(t) + C\underline{y}^3(t) = 8, \quad 0 \le t \le 1, \tag{14.16}$$

$$A\overline{y}''(t) + BD^{3/2}\overline{y}(t) + C\overline{y}^3(t) = 8, \quad 0 \le t \le 1. \tag{14.17}$$

By the homotopy technique, we construct a homotopy for Eqs. (14.16) and (14.17) as follows:

$$H\left(\underline{Y},p\right) = (1-p)\left[\frac{d^2\underline{Y}}{d^2t} - \frac{d^2\underline{y}_0}{d^2t}\right] + p\left[\frac{d^2\underline{Y}}{d^2t} + \frac{B}{A}\frac{d^{3/2}\underline{Y}(t)}{dt^{3/2}} + \frac{C}{A}\underline{Y}^3(t) - \frac{8}{A}\right] = 0,$$

$$(14.18)$$

$$H\left(\overline{Y},p\right) = (1-p)\left[\frac{d^2\overline{Y}}{d^2t} - \frac{d^2\overline{y}}{d^2t}\right] + p\left[\frac{d^2\overline{Y}}{d^2t} + \frac{B}{A}\frac{d^{3/2}\overline{Y}(t)}{dt^{3/2}} + \frac{C}{A}\overline{Y}^3(t) - \frac{8}{A}\right] = 0.$$

$$(14.19)$$

One may try to obtain a solution of Eqs. (14.16) and (14.17) in the form

$$\underline{Y}(t) = \underline{Y}_0(t) + p\underline{Y}_1(t) + p^2\underline{Y}_2(t) + \cdots, \tag{14.20}$$

$$\overline{Y}(t) = \overline{Y}_0(t) + p\overline{Y}_1(t) + p^2\overline{Y}_2(t) + \cdots \tag{14.21}$$

where, $\underline{Y}_i(t)$ and $\overline{Y}_i(t)$ for $i = 0, 1, 2, \ldots$ are functions yet to be determined. Substituting Eqs. (14.20) and (14.21) into Eqs. (14.16) and (14.17), respectively, and equating the terms with the identical powers of p (as done previously), we have

$$p^0 : \begin{cases} \dfrac{d^2\underline{Y}_0}{dt^2} - \dfrac{d^2\underline{y}_0}{dt^2} = 0, \\[2mm] \dfrac{d^2\overline{Y}_0}{dt^2} - \dfrac{d^2\overline{y}_0}{dt^2} = 0, \end{cases}$$

$$p^1 : \begin{cases} \dfrac{d^2\underline{Y}_1}{dt^2} + \dfrac{d^2\underline{y}_0}{dt^2} + \dfrac{B}{A}\dfrac{d^{3/2}\underline{Y}_0}{dt^{3/2}} + \dfrac{C}{A}\underline{Y}_0^3 - \dfrac{8}{A} = 0, \\[2mm] \dfrac{d^2\overline{Y}_1}{dt^2} + \dfrac{d^2\overline{y}_0}{dt^2} + \dfrac{B}{A}\dfrac{d^{3/2}\overline{Y}_0}{dt^{3/2}} + \dfrac{C}{A}\overline{Y}_0^3 - \dfrac{8}{A} = 0, \end{cases}$$

$$p^2 : \begin{cases} \dfrac{d^2\underline{Y}_2}{dt^2} + \dfrac{B}{A}\dfrac{d^{3/2}\underline{Y}_1}{dt^{3/2}} + \dfrac{3C}{A}\underline{Y}_0^2\underline{Y}_1 = 0, \\[2mm] \dfrac{d^2\overline{Y}_2}{dt^2} + \dfrac{B}{A}\dfrac{d^{3/2}\overline{Y}_1}{dt^{3/2}} + \dfrac{3C}{A}\overline{Y}_0^2\overline{Y}_1 = 0, \end{cases}$$

and so on.

As in the previous cases, the initial approximation $\underline{Y}_0(t;r)$ or $\underline{y}_0(t;r)$ and $\overline{Y}_1(t;r)$ or $\overline{y}_0(t;r)$ can be freely chosen. Here we set

$$\underline{Y}_0(t;r) = \underline{y}_0(t;r) = 0.1r - 0.1 \quad \text{and} \quad \overline{Y}_0(t;r) = \overline{y}_0(t;r) = 0.1 - 0.1r.$$

Substituting the assumed initial condition into p^i terms for $i = 0, 1, 2, \ldots$, respectively, yields

$$\underline{Y}_1(t; r) = -\frac{C}{A}(0.1r - 0.1)^3 \frac{t^2}{\Gamma(3)} + \frac{8t^2}{A\Gamma(3)},$$

$$\overline{Y}_1(t; r) = -\frac{C}{A}(0.1 - 0.1r)^3 \frac{t^2}{\Gamma(3)} + \frac{8t^2}{A\Gamma(3)},$$

$$\underline{Y}_2(t; r) = \frac{BC}{A^2}(0.1r - 0.1)^3 \frac{t^{5/2}}{\Gamma(7/2)} - \frac{8Bt^{5/2}}{A^2\Gamma(7/2)} + (0.1r - 0.1)^5 \left(\frac{3C^2}{A^2}\right) \frac{t^4}{\Gamma(5)}$$

$$- (0.1r - 0.1) \left(\frac{24C}{A^2}\right) \frac{t^4}{\Gamma(5)},$$

$$\overline{Y}_2(t) = \frac{BC}{A^2}(0.1 - 0.1r)^3 \frac{t^{5/2}}{\Gamma(7/2)} - \frac{8Bt^{5/2}}{A^2\Gamma(7/2)} + (0.1 - 0.1r)^5 \left(\frac{3C^2}{A^2}\right) \frac{t^4}{\Gamma(5)}$$

$$- (0.1 - 0.1r) \left(\frac{24C}{A^2}\right) \frac{t^4}{\Gamma(5)},$$

and so on. In the same manner, the rest of components can also be obtained. The third-term approximate solution for Eqs. (14.16) and (14.17) may be obtained as

$$\underline{y}(t; r) \cong \underline{Y}_0(t; r) + \underline{Y}_1(t; r) + \underline{Y}_2(t; r)$$

$$= (0.1r - 0.1) - \frac{C}{A}(0.1r - 0.1)^3 \frac{t^2}{\Gamma(3)} + \frac{8t^2}{A\Gamma(3)} + \frac{BC}{A^2}(0.1r - 0.1)^3 \frac{t^{5/2}}{\Gamma(7/2)}$$

$$- \frac{8Bt^{5/2}}{A^2\Gamma(7/2)} + (0.1r - 0.1)^5 \left(\frac{3C^2}{A^2}\right) \frac{t^4}{\Gamma(5)} - (0.1r - 0.1) \left(\frac{24C}{A^2}\right) \frac{t^4}{\Gamma(5)},$$

$$\overline{y}(t; r) \cong \overline{Y}_0(t; r) + \overline{Y}_1(t; r) + \overline{Y}_2(t; r)$$

$$= (0.1 - 0.1r) - \frac{C}{A}(0.1 - 0.1r)^3 \frac{t^2}{\Gamma(3)} + \frac{8t^2}{A\Gamma(3)} + \frac{BC}{A^2}(0.1 - 0.1r)^3 \frac{t^{5/2}}{\Gamma(7/2)}$$

$$- \frac{8Bt^{5/2}}{A^2\Gamma(7/2)} + (0.1 - 0.1r)^5 \left(\frac{3C^2}{A^2}\right) \frac{t^4}{\Gamma(5)} - (0.1 - 0.1r) \left(\frac{24C}{A^2}\right) \frac{t^4}{\Gamma(5)}.$$

14.2 RESULTS AND DISCUSSIONS

In this section, numerical solutions of linear and nonlinear fuzzy fractional Bagley–Torvik equations using HPM for different cases as discussed earlier have been presented. Few representative results with respect to various parameters involved in the corresponding equation of three cases are reported as follows.

- In the first case, linear inhomogeneous fuzzy Bagley–Torvik equation for $\alpha = 3/2, A = 1, B = 0.5$, and $C = 0.5$ with $f(t) = 8$ for $0 \le t \le 1$ is solved by

HPM, and the solutions are given in Tables 14.1(a) and 14.1(b) for different values of r and t.

• For the second case, the same simulation has been performed using the same parameters as in first case with $f(t) = (1 + t)$. The obtained results are given in Tables 14.2(a) and 14.2(b).

• Lastly, in the third case, nonlinear inhomogeneous fuzzy Bagley–Torvik equation is solved by HPM for $\alpha = 3/2, A = 1, B = 1$, and $C = 1$ with $f(t) = 8$. Computed solutions are tabulated for various values of r and t and are incorporated in Tables 14.3(a) and 14.3(b).

TABLE 14.1(a) HPM Solution of Case 1 for $t = 0, t = 0.5$, and $t = 1$

r	$[\underline{y}(t,r),\ \overline{y}(t,r)]$		
	$t = 0$	$t = 0.5$	$t = 1$
0	[−0.1, 0.1]	[0.7288, 0.9149]	[2.9758, 3.1086]
0.1	[−0.09, 0.09]	[0.7381, 0.9056]	[2.9825, 3.1020]
0.2	[−0.08, 0.08]	[0.7474, 0.8963]	[2.9891, 3.0953]
0.3	[−0.07, 0.07]	[0.7567, 0.8870]	[2.9957, 3.0887]
0.4	[−0.06, 0.06]	[0.7660, 0.8777]	[3.0024, 3.0821]
0.5	[−0.05, 0.05]	[0.7753, 0.8684]	[3.0090, 3.0754]
0.6	[−0.04, 0.04]	[0.7846, 0.8591]	[3.0157, 3.0688]
0.7	[−0.03, 0.03]	[0.7939, 0.8498]	[3.0223, 3.0621]
0.8	[−0.02, 0.02]	[0.8032, 0.8405]	[3.0289, 3.0555]
0.9	[−0.01, 0.01]	[0.8125, 0.8312]	[3.0356, 3.0489]
1	[0, 0]	[0.8219, 0.8219]	[3.0422, 3.0422]

TABLE 14.1(b) HPM Solution of Case 1 for $r = 0, r = 0.5$, and $r = 1$

t	$[\underline{y}(t,r),\ \overline{y}(t,r)]$		
	$r = 0$	$r = 0.5$	$r = 1$
0	[−0.1, 0.1]	[−0.05, 0.05]	[0, 0]
0.1	[−0.0632, 0.1362]	[−0.0133, 0.0863]	[0.0365, 0.0365]
0.2	[0.04196, 0.23997]	[0.0914, 0.1904]	[0.1409, 0.1409]
0.3	[0.2109, 0.4063]	[0.2598, 0.3574]	[0.3086, 0.3086]
0.4	[0.4408, 0.63223]	[0.4886, 0.5843]	[0.5365, 0.5365]
0.5	[0.7288, 0.9149]	[0.7753, 0.8684]	[0.8219, 0.8219]
0.6	[1.0729, 1.2521]	[1.1177, 1.2073]	[1.1625, 1.1625]
0.7	[1.4712, 1.6418]	[1.5139, 1.5992]	[1.5565, 1.5565]
0.8	[1.9221, 2.0821]	[1.9621, 2.0421]	[2.0021, 2.0021]
0.9	[2.4240, 2.5715]	[2.4609, 2.5346]	[2.4978, 2.4978]
1	[2.9758, 3.1086]	[3.0090, 3.0754]	[3.0422, 3.0422]

TABLE 14.2(a)   HPM Solution of Case 2 for $t = 0$, $t = 0.5$, and $t = 1$

r	$[\underline{y}(t,r),\ \overline{y}(t,r)]$		
	$t = 0$	$t = 0.5$	$t = 1$
0	[−0.1, 0.1]	[0.0166552, 0.198306]	[0.421241, 0.548502]
0.1	[−0.09, 0.09]	[0.0257377, 0.189223]	[0.427604, 0.542139]
0.2	[−0.08, 0.08]	[0.0348202, 0.180141]	[0.433967, 0.535776]
0.3	[−0.07, 0.07]	[0.0439028, 0.171058]	[0.44033, 0.529413]
0.4	[−0.06, 0.06]	[0.0529853, 0.161976]	[0.446693, 0.523049]
0.5	[−0.05, 0.05]	[0.0620678, 0.152893]	[0.453056, 0.516686]
0.6	[−0.04, 0.04]	[0.0711503, 0.143811]	[0.459419, 0.510323]
0.7	[−0.03, 0.03]	[0.0802329, 0.134728]	[0.465782, 0.50396]
0.8	[−0.02, 0.02]	[0.0893154, 0.125646]	[0.472145, 0.497597]
0.9	[−0.01, 0.01]	[0.0983979, 0.116563]	[0.478508, 0.491234]
1	[0, 0]	[0.10748, 0.10748]	[0.484871, 0.484871]

TABLE 14.2(b)   HPM Solution of Case 2 for $r = 0$, $r = 0.5$, and $r = 1$

t	$[\underline{y}(t,r),\ \overline{y}(t,r)]$		
	$r = 0$	$r = 0.5$	$r = 1$
0	[−0.1, 0.1]	[−0.05, 0.05]	[0, 0]
0.1	[−0.0952, 0.1039]	[−0.0454, 0.0541]	[0.0043, 0.0043]
0.2	[−0.0814, 0.1154]	[−0.0322, 0.0662]	[0.0170, 0.0170]
0.3	[−0.0584, 0.1346]	[−0.0102, 0.0863]	[0.0380552, 0.0380552]
0.4	[−0.0260, 0.1620]	[0.0209, 0.115]	[0.0679887, 0.0679887]
0.5	[0.0166, 0.198306]	[0.0620678, 0.152893]	[0.10748, 0.10748]
0.6	[0.070444, 0.244325]	[0.113914, 0.200855]	[0.157385, 0.157385]
0.7	[0.136399, 0.301071]	[0.177567, 0.259903]	[0.218735, 0.218735
0.8	[0.215798, 0.369722]	[0.254279, 0.331241]	[0.29276, 0.29276]
0.9	[0.310153, 0.451661]	[0.34553, 0.416284]	[0.380907, 0.380907]
1	[0.421241, 0.548502]	[0.453056, 0.516686]	[0.484871, 0.484871]

One may notice that for $r = 1$, the fuzzy initial condition converts into crisp initial value, and also for other particular value of r, it converts into interval initial value problem. Triangular fuzzy and interval solutions are depicted for Case 1 in Fig. 14.1(a) and (b), respectively. Next, Fig. 14.2(a) and (b), respectively, represent the fuzzy and interval results for Case 2. As such, Fig. 14.3(a) and (b), respectively, show the fuzzy and interval results for Case 3.

Figures 14.1(a), 14.2(a), and 14.3(a) depict triangular fuzzy solutions of all the three cases for $t = 0$, 0.5, and 1. Similarly, Figs. 14.1(b), 14.2(b), and 14.3(b) give interval solution bounds for all the three cases for $r = 0$, 0.5, and 1. Here, thick and dashed lines correspond to lower and upper bounds of the fuzzy and interval solutions, respectively.

TABLE 14.3(a)   HPM Solution of Case 3 for $t = 0$, $t = 0.5$, and $t = 1$

r	$[\underline{y}(t,r),\ \overline{y}(t,r)]$		
	$t = 0$	$t = 0.5$	$t = 1$
0	[−0.1, 0.1]	[0.685651, 0.888186]	[2.68412, 2.89867]
0.1	[−0.09, 0.09]	[0.695826, 0.878129]	[2.69576, 2.88893]
0.2	[−0.08, 0.08]	[0.705999, 0.868063]	[2.70731, 2.87909]
0.3	[−0.07, 0.07]	[0.716168, 0.857987]	[2.71877, 2.86913]
0.4	[−0.06, 0.06]	[0.726334, 0.847903]	[2.73014, 2.85906]
0.5	[−0.05, 0.05]	[0.736496, 0.83781]	[2.74142, 2.84888]
0.6	[−0.04, 0.04]	[0.746653, 0.827709]	[2.7526, 2.83859]
0.7	[−0.03, 0.03]	[0.756805, 0.8176]	[2.7637, 2.8282]
0.8	[−0.02, 0.02]	[0.766953, 0.807484]	[2.77469, 2.8177]
0.9	[−0.01, 0.01]	[0.777095, 0.797361]	[2.78559, 2.8071]
1	[0, 0]	[0.787231, 0.787231]	[2.7964, 2.7964]

TABLE 14.3(b)   HPM Solution of Case 3 for $r = 0$, $r = 0.5$, and $r = 1$

t	$[\underline{y}(t,r),\ \overline{y}(t,r)]$		
	$r = 0$	$r = 0.5$	$r = 1$
0	[−0.1, 0.1]	[−0.05, 0.05]	[0, 0]
0.1	[−0.0638279, 0.136215]	[−0.0138178, 0.086205]	[0.0361939, 0.036193]
0.2	[0.0383367, 0.238586]	[0.0884012, 0.188533]	[0.138469, 0.138469]
0.3	[0.200279, 0.400976]	[0.250476, 0.350841]	[0.300668, 0.300668]
0.4	[0.417355, 0.618797]	[0.467796, 0.568547]	[0.518204, 0.518204]
0.5	[0.685651, 0.888186]	[0.736496, 0.83781]	[0.787231, 0.787231]
0.6	[1.00171, 1.20573]	[1.05317, 1.15524]	[1.10437, 1.10437]
0.7	[1.3624, 1.56833]	[1.41474, 1.51779]	[1.46657, 1.46657]
0.8	[1.76482, 1.97311]	[1.81837, 1.92264]	[1.87102, 1.87102]
0.9	[2.20625, 2.41741]	[2.26142, 2.36715]	[2.31511, 2.31511]
1	[2.68412, 2.89867]	[2.74142, 2.84888]	[2.7964, 2.7964]

As discussed earlier for $r = 1$, fuzzy initial condition converts into crisp initial value. As such, for crisp initial value, the obtained results by HPM are compared with the numerical solution of Podlubny (1999) with the step size $h = 0.0005$ for all the three cases. Computed results are tabulated in Table 14.4, and these are found to be in good agreement. The accuracy can be improved for all the cases by computing more number of terms in HPM.

It is interesting to note that for all the cases, the lower and upper bounds are the same for $r = 1$ and also approximately equal to those of the crisp solution obtained by Podlubny (1999).

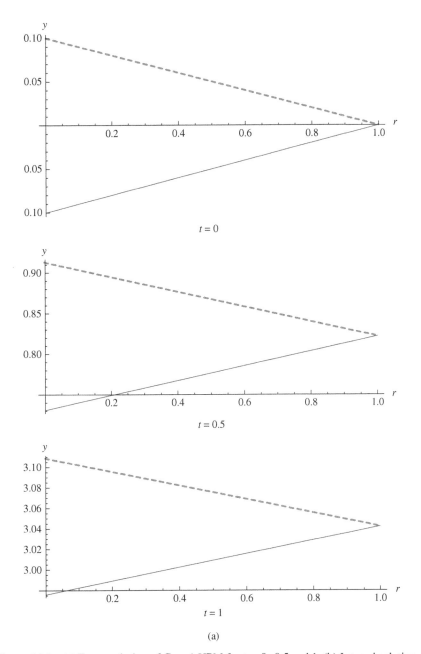

$t = 0$

$t = 0.5$

$t = 1$

(a)

Figure 14.1 (a) Fuzzy solution of Case 1 HPM for $t = 0$, 0.5 and 1. (b) Interval solution of Case 1 HPM for $r = 0, 0.5$, and 1

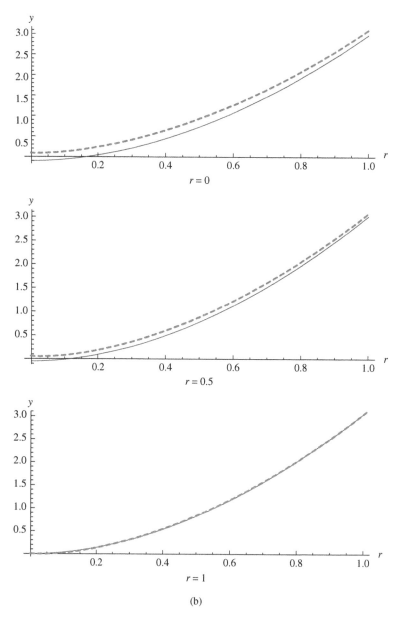

$r = 0$

$r = 0.5$

$r = 1$

(b)

Figure 14.1 (*Continued*)

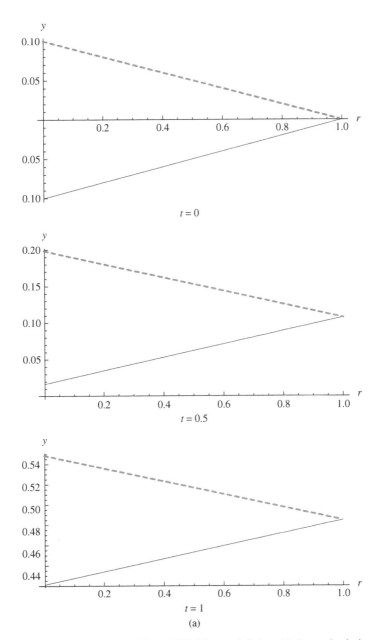

Figure 14.2 (a) Fuzzy solution of Case 2 HPM for $t = 0, 0.5$, and 1. Interval solution of Case 2 HPM for $r = 0, 0.5$, and 1

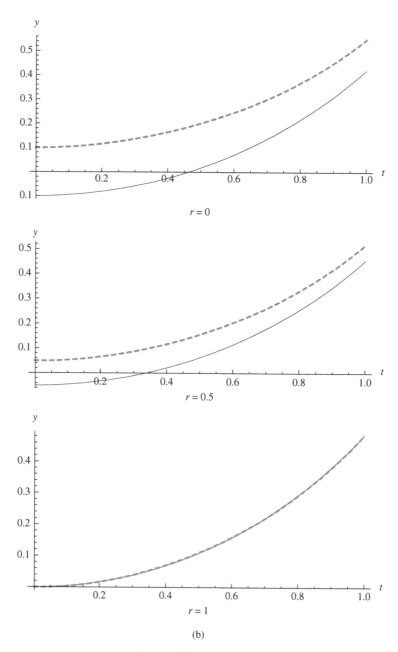

Figure 14.2 (*Continued*)

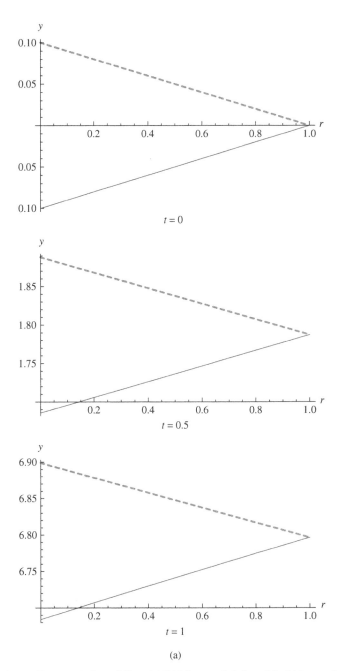

(a)

Figure 14.3   (a) Fuzzy solution of Case 3 HPM for $t = 0, 0.5$, and 1. (b) Interval solution of Case 3 HPM for $r = 0, 0.5$, and 1

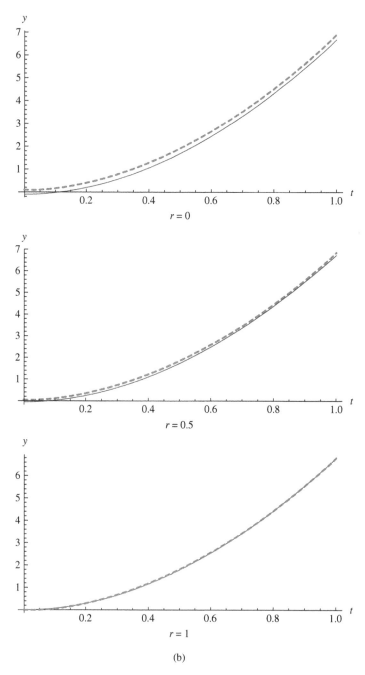

(b)

Figure 14.3 (*Continued*)

TABLE 14.4   Comparison Between HPM Solution and Podlubny (1999)

t	0	0.2	0.4	0.6	0.8	1
Case 1						
Present ($r = 1$)	0	0.1409	0.5365	1.1625	2.0021	3.0422
Podlubny (1999)	0	0.1402	0.5325	1.1478	1.9617	2.9510
Absolute error	0	0.0007	0.0040	0.0147	0.0404	0.0912
Case 2						
Present ($r = 1$)	0	0.0170022	0.0679887	0.157385	0.29276	0.484871
Podlubny (1999)	0	0.0166974	0.064838	0.14397	0.254001	0.394513
Absolute error	0	0.0003048	0.0031507	0.013415	0.038759	0.090358
Case 3						
Present ($r = 1$)	0	0.138469	0.518204	1.10437	1.87102	2.7964
Podlubny (1999)	0	0.140491	0.535605	1.15846	1.96543	2.83719
Absolute error	0	0.002022	0.017401	0.05409	0.09441	0.04079

In this chapter, HPM has been successfully applied to find the solution of fuzzy fractional Bagley–Torvik linear and nonlinear differential equations. The solution obtained by HPM is an infinite series with appropriate initial condition, which in turn may be expressed in a closed form that refers to the exact solution. The result shows that the HPM is a powerful mathematical tool to solve fuzzy fractional Bagley–Torvik linear and nonlinear differential equations.

BIBLIOGRAPHY

Abdulaziz O, Hashima I, Momani S. Solving systems of fractional differential equations by homotopy-perturbation method. *Phys Lett A* 2008;**372**:451–459.

Agrawal RP, Lakshmikantham V, Nieto JJ. On the concept of solution for fractional differential equations with uncertainty. *Nonlinear Anal* 2010;**72**:2859–2862.

Allahviranloo T, Salahshour S, Abbasbandy S. Explicit solutions of fractional differential equations with uncertainty. *Soft Comput* 2012;**16**:297–302.

Arshad S, Lupulescu V. On the fractional differential equations with uncertainty. *Nonlinear Anal* 2011a;**74**:3685–3693.

Arshad S, Lupulescu V. Fractional differential equation with the fuzzy initial condition. *Electron J Differ Eq* 2011b;**2011**:1–8.

Bagley RL, Torvik PJ. On the appearance of the fractional derivative in the behaviour of real materials. *Appl Mech* 1984;**51**:294–298.

Bede B. Note on numerical solutions of fuzzy differential equations by predictor-corrector method. *Inform Sci* 2008;**178**:1917–1922.

Cenesiz Y, Keskin Y, Kurnaz A. The solution of the Bagley–Torvik equation with the generalized Taylor collocation method. *J Franklin Inst* 2010;**347**:452–466.

EL-Sayed AMA, EL-Mesiry AEM, EL-Saka HAA. Numerical solution for multi-term fractional (arbitrary) orders differential equations. *Comput Appl Math* 2004;**23**:33–54.

Gejji VD, Jafari H. Adomian decomposition: a tool for solving a system of fractional differential equations. *J Math Anal Appl* 2005;**301**:508–518.

Gejji VD, Jafari H. Solving a multi-order fractional differential equation using Adomian decomposition. *Appl Math Comput* 2007;**189**:541–548.

He JH. Nonlinear oscillation with fractional derivative and its applications. Proceedings of the International Conference on Vibrating Engineering; Dalian, China; 1998. pp. 288–291.

He JH. Homotopy perturbation technique. *Comp Methods Appl Mech Eng* 1999;**178**:257–262.

He JH. A coupling method of homotopy technique and a perturbation technique for non-linear problems. *Int J Nonlinear Mech* 2000;**35**:37–43.

Hu Y, Luo Y, Lu Z. Analytical solution of the linear fractional differential equation by Adomian decomposition method. *J Comput Appl Math* 2008;**215**:220–229.

Jafari H, Momani S. Solving fractional diffusion and wave equations by modified homotopy perturbation method. *Phys Lett A* 2007;**370**:388–396.

Kilbas AA, Srivastava HM, Trujillo JJ. *Theory and Applications of Fractional Differential Equations*. Amsterdam: Elsevier Sciences B.V.; 2006.

Kiryakova V. *Generalized Fractional Calculus and Applications*. New York: Longman Scientific & Technical, Harlow, copublished in the United States with John Wiley & Sons Inc.; 1994.

Lakshmikantham V, Mohapatra RN. *Theory of Fuzzy Differential Equations and Applications*. London: Taylor & Francis; 2003.

Miller KS, Ross B. *An Introduction to the Fractional Calculus and Fractional Differential Equations*. New York: John Wiley & Sons; 1993.

Mohammed OH, Fadhel FS, Abdul- Khaleq FA. Differential Transform method for solving fuzzy fractional initial value problems. *J Basrah Res (Sci)* 2011;**37**:158–170.

Momani S, Odibat Z. Numerical comparison of methods for solving linear differential equations of fractional order. *Chaos Solitons Fract* 2007;**31**:1248–1255.

Podlubny I. *Fractional Differential Equations*. San Diego: Academic Press; 1999.

Salahshour S, Allahviranloo T, Abbasbandy S. Solving fuzzy fractional differential equations by fuzzy Laplace transforms. *Commun Nonlinear Sci Numer Simulat* 2012;**17**:1372–1381.

Ross TJ. *Fuzzy Logic with Engineering Applications*. Wiley Student Edition; 2007.

Trinks C, Ruge P. Treatment of dynamic systems with fractional derivatives without evaluating memory-integrals. *Comput Mech* 2002;**29**:471–476.

Wang H, Liu Y. Existence results for fractional fuzzy differential equations with finite delay. *Int Math Forum* 2011;**6**:2535–2538.

Zimmermann HJ. *Fuzzy Set Theory and Its Application*. Vol. **2001**. Boston/Dordrecht/London: Kluwer Academic Publishers; 2001.

APPENDIX A

Crisp fractional differential equations are well known, and there exist various methods to solve them. But this appendix is included to have straightforward understanding of solution method(s) of crisp fractionally differential equation for the sake of completeness. As such, the readers may easily understand the fuzzy fractionally differential equation, which is the main aim of this book. Here, fractionally damped spring–mass system and beam problems are considered, respectively, as example problems, and those are discussed next as Problems 1 and 2 with crisp parameters and initial conditions. Homotopy perturbation method (HPM) has been used for the solution process.

A.1 FRACTIONALLY DAMPED SPRING–MASS SYSTEM (PROBLEM 1)

We aim here to estimate the dynamic response of a fractionally (crisp) damped discrete system. Accordingly, let us consider a single degree-of-freedom spring–mass–damper system (Suarez and Shokooh, 1997; Yuan and Agrawal, 2002), which may be described by the following differential equation

$$mD^2x(t) + cD^\alpha x(t) + kx(t) = f(t) \tag{A.1}$$

where, m, c, and k represent the mass, damping, and stiffness coefficients, respectively. $f(t)$ is the externally applied force, and $D^\alpha x(t), 0 < \alpha < 1$, is the derivative of

Fuzzy Arbitrary Order System: Fuzzy Fractional Differential Equations and Applications, First Edition.
Snehashish Chakraverty, Smita Tapaswini, and Diptiranjan Behera.
© 2016 John Wiley & Sons, Inc. Published 2016 by John Wiley & Sons, Inc.

order α of the displacement function $x(t)$. Although the coefficient α (known as the memory parameter) may take any value between 0 and 1, but here the value has been adopted as 1/2. It has been shown that it describes the frequency dependence of the damping materials quite satisfactorily in the crisp fractional dynamic systems (Suarez and Shokooh, 1997; Yuan and Agrawal, 2002). The initial conditions are considered as the initial displacement $x(0) = 0$ and the initial velocity $v(0) = \dot{x}(0) = 0$.

Now Eq. (A.1) can be written as

$$D^2 x(t) + \frac{c}{m} D^{1/2} x(t) + \frac{k}{m} x(t) = \frac{f(t)}{m}. \tag{A.2}$$

According to HPM, one may construct a simple homotopy for an embedding parameter $p \in [0, 1]$ as follows:

$$(1 - p)D^2 x(t) + p\left(D^2 x(t) + \frac{c}{m} D^{1/2} x(t) + \frac{k}{m} x(t) - \frac{f(t)}{m}\right) = 0, \tag{A.3}$$

or

$$D^2 x(t) + p\left(\frac{c}{m} D^{1/2} x(t) + \frac{k}{m} x(t) - \frac{f(t)}{m}\right) = 0. \tag{A.4}$$

As discussed earlier, p is considered as a small homotopy parameter $0 \le p \le 1$. So in the changing process from 0 to 1, for $p = 0$, Eqs. (A.3) and (A.4) become a linear equation, that is, $D^2 x(t) = 0$, which is easy to solve. For $p = 1$, Eqs. (A.3) and (A.4) turn out to be the same as the original equation (A.1) or (A.2). This is called deformation in topology. $D^2 x(t)$ and $\frac{c}{m} D^{1/2} x(t) + \frac{k}{m} x(t) - \frac{f(t)}{m}$ are called homotopic.

Assuming the solution of Eq. (A.3) or (A.4) as a power series expansion in p as

$$x(t) = x_0(t) + p x_1(t) + p^2 x_2(t) + p^3 x_3(t) + \cdots, \tag{A.5}$$

where $x_i(t), i = 0, 1, 2, \ldots$ are functions yet to be determined. Substituting Eq. (A.5) into Eq. (A.3) or (A.4), and equating the terms with the identical power of p, we can obtain a series of equations of the form

$$p^0 : D^2 x_0(t) = 0,$$

$$p^1 : D^2 x_1(t) + \frac{c}{m} D^{1/2} x_0(t) + \frac{k}{m} x_0(t) - \frac{f(t)}{m} = 0,$$

$$p^2 : D^2 x_2(t) + \frac{c}{m} D^{1/2} x_1(t) + \frac{k}{m} x_1(t) = 0,$$

$$p^3 : D^2 x_3(t) + \frac{c}{m} D^{1/2} x_2(t) + \frac{k}{m} x_2(t) = 0, \tag{A.6}$$

$$p^4 : D^2 x_4(t) + \frac{c}{m} D^{1/2} x_3(t) + \frac{k}{m} x_3(t) = 0,$$

$$p^5 : D^2 x_5(t) + \frac{c}{m} D^{1/2} x_4(t) + \frac{k}{m} x_4(t) = 0,$$

$$p^6 : D^2 x_6(t) + \frac{c}{m} D^{1/2} x_5(t) + \frac{k}{m} x_5(t) = 0,$$

and so on.

Applying the operator L_{tt}^{-1} (which is the inverse of the operator $L_{tt} = D^2$) on both sides of Eq. (A.6), one may obtain the following equations:

$$x_0(t) = 0,$$

$$x_1(t) = L_{tt}^{-1} \left(-\frac{c}{m} D^{1/2} x_0(t) - \frac{k}{m} x_0(t) + \frac{f(t)}{m} \right)$$

$$= \frac{c}{m} D^{-3/2} x_0(t) - \frac{k}{m} D^{-2} x_0(t) + D^{-2} \frac{f(t)}{m},$$

$$x_2(t) = L_{tt}^{-1} \left(-\frac{c}{m} D^{1/2} x_1(t) - \frac{k}{m} x_1(t) \right)$$

$$= -\frac{c}{m} D^{-3/2} x_1(t) - \frac{k}{m} D^{-2} x_1(t),$$

$$x_3(t) = L_{tt}^{-1} \left(-\frac{c}{m} D^{1/2} x_2(t) - \frac{k}{m} x_2(t) \right)$$

$$= -\frac{c}{m} D^{-3/2} x_2(t) - \frac{k}{m} D^{-2} x_2(t), \qquad\qquad (A.7)$$

$$x_4(t) = L_{tt}^{-1} \left(-\frac{c}{m} D^{1/2} x_3(t) - \frac{k}{m} x_3(t) \right)$$

$$= -\frac{c}{m} D^{-3/2} x_3(t) - \frac{k}{m} D^{-2} x_3(t),$$

$$x_5(t) = L_{tt}^{-1} \left(-\frac{c}{m} D^{1/2} x_4(t) - \frac{k}{m} x_4(t) \right)$$

$$= -\frac{c}{m} D^{-3/2} x_4(t) - \frac{k}{m} D^{-2} x_4(t),$$

$$x_6(t) = L_{tt}^{-1} \left(-\frac{c}{m} D^{1/2} x_5(t) - \frac{k}{m} x_5(t) \right)$$

$$= -\frac{c}{m} D^{-3/2} x_5(t) - \frac{k}{m} D^{-2} x_5(t),$$

and so on.

Now substituting these terms in Eq. (A.5), one may get the approximate solution of Eq. (A.1) as

$$x(t) = x_0(t) + x_1(t) + x_2(t) + x_3(t) + x_4(t) + x_5(t) + x_6(t) + \cdots$$

The solution series converge very rapidly. The proof of convergence of the afore-mentioned series may be found in He (1999, 2000). The rapid convergence means that only few terms are sufficient to get the approximate solutions.

A.1.1 Response Analysis

In this section, the response of fractionally damped spring–mass system has been analyzed with respect to unit step and impulse loading.

A.1.1.1 Step Function Response   Let us consider a stationary oscillator subject to an excitation of the form $f(t) = u(t)$, where $u(t)$ is the Heaviside function with unit step load in Eq. (A.1). By using HPM, we have

$$x_0(t) = 0,$$

$$x_1(t) = \frac{1}{2m} t^2 u(t),$$

$$x_2(t) = \left(-\frac{c}{m^2} \frac{t^{7/2}}{\Gamma(9/2)} - \frac{k}{m^2} \frac{t^4}{\Gamma(5)} \right) u(t),$$

$$x_3(t) = \left(\frac{c^2}{m^3} \frac{t^5}{\Gamma(6)} + \frac{2kc}{m^3} \frac{t^{11/2}}{\Gamma(13/2)} + \frac{k^2}{m^3} \frac{t^6}{\Gamma(7)} \right) u(t), \tag{A.8}$$

$$x_4(t) = \left(-\frac{c^3}{m^4} \frac{t^{13/2}}{\Gamma(15/2)} - \frac{3kc^2}{m^4} \frac{t^7}{\Gamma(8)} - \frac{3k^2 c}{m^4} \frac{t^{15/2}}{\Gamma(17/2)} - \frac{k^3}{m^4} \frac{t^8}{\Gamma(9)} \right) u(t),$$

$$x_5(t) = \left(\frac{c^4}{m^5} \frac{t^8}{\Gamma(9)} + \frac{4kc^3}{m^5} \frac{t^{17/2}}{\Gamma(19/2)} + \frac{6k^2 c^2}{m^5} \frac{t^9}{\Gamma(10)} \right.$$
$$\left. + \frac{4k^3 c}{m^5} \frac{t^{19/2}}{\Gamma(21/2)} + \frac{k^4}{m^5} \frac{t^{10}}{\Gamma(11)} \right) u(t),$$

and so on.

In a similar manner, the rest of the components can be obtained. Therefore, the solution can be written in general form as

$$x(t) = \frac{u(t)}{m} \sum_{r=0}^{\infty} \frac{(-1)^r}{r!} \left(\frac{k}{m} \right)^r t^{2(r+1)} \sum_{j=0}^{\infty} \left(\frac{-c}{m} \right)^j \frac{(j+r)! t^{3j/2}}{j! \Gamma\left(\frac{3j}{2} + 2r + 3 \right)} \tag{A.9}$$

$$x(t) = \frac{u(t)}{m} \sum_{r=0}^{\infty} \frac{(-1)^r}{r!} \left(\frac{k}{m} \right)^r t^{2(r+1)} E_{3/2, r/2+3}^r \left(\frac{-c}{m} t^{3/2} \right). \tag{A.10}$$

Now Eq. (A.10) can be rewritten as

$$x(t) = \frac{u(t)}{m} \sum_{r=0}^{\infty} \frac{(-1)^r}{r!} \left(\omega_n^2 \right)^r t^{2(r+1)} E_{3/2, r/2+3}^r \left(-2\eta \omega_n^{3/2} t^{3/2} \right)$$

where $\omega_n = \sqrt{k/m}$ and $\eta = c/2m\omega_n^{3/2}$ are the natural frequency and damping ratio, respectively.

A.1.1.2 Impulse Function Response Next, let us consider the response subject to a unit impulse load, that is, $f(t) = \delta(t)$, where $\delta(t)$ is the unit impulse function. Again by using HPM, we obtain

$$x_0(t) = 0,$$

$$x_1(t) = \frac{t}{m},$$

$$x_2(t) = -\frac{c}{m^2}\frac{t^{5/2}}{\Gamma(7/2)} - \frac{k}{m^2}\frac{t^3}{\Gamma(4)},$$

$$x_3(t) = \frac{c^2}{m^3}\frac{t^4}{\Gamma(5)} + \frac{2kc}{m^3}\frac{t^{9/2}}{\Gamma(11/2)} + \frac{k^2}{m^3}\frac{t^5}{\Gamma(6)}, \tag{A.11}$$

$$x_4(t) = -\frac{c^3}{m^4}\frac{t^{11/2}}{\Gamma(13/2)} - \frac{3kc^2}{m^4}\frac{t^6}{\Gamma(7)} - \frac{3k^2c}{m^4}\frac{t^{13/2}}{\Gamma(15/2)} - \frac{k^3}{m^4}\frac{t^7}{\Gamma(8)},$$

$$x_5(t) = \frac{c^4}{m^5}\frac{t^7}{\Gamma(8)} + \frac{4kc^3}{m^5}\frac{t^{15/2}}{\Gamma(17/2)} + \frac{6k^2c^2}{m^5}\frac{t^8}{\Gamma(9)} + \frac{4k^3c}{m^5}\frac{t^{17/2}}{\Gamma(19/2)} + \frac{k^4}{m^5}\frac{t^9}{\Gamma(10)},$$

and so on.

Accordingly, the general solution may be written as

$$x(t) = \frac{1}{m}\sum_{r=0}^{\infty}\frac{(-1)^r}{r!}\left(\frac{k}{m}\right)^r t^{2r+1}\sum_{j=0}^{\infty}\left(\frac{-c}{m}\right)^j\frac{(j+r)!t^{3j/2}}{j!\Gamma\left(\frac{3j}{2}+2r+2\right)} \tag{A.12}$$

$$= \frac{1}{m}\sum_{r=0}^{\infty}\frac{(-1)^r}{r!}\left(\frac{k}{m}\right)^r t^{2r+1}E_{3/2,r/2+2}^r\left(\frac{-c}{m}t^{3/2}\right). \tag{A.13}$$

Substituting $\omega_n = \sqrt{k/m}$ and $\eta = c/2m\omega_n^{3/2}$ in Eq. (A.12), it gives

$$x(t) = \frac{1}{m}\sum_{r=0}^{\infty}\frac{(-1)^r}{r!}\left(\omega_n^2\right)^r t^{2r+1}E_{3/2,r/2+2}^r\left(-2\eta\omega_n^{3/2}t^{3/2}\right).$$

A.1.2 Analytical Solution Using Fractional Green's Function

Analytical solution of Eq. (A.1) can be obtained by using fractional Green's function for a three-term fractional differential equation with constant coefficients (Section 5.4 of Podlubny (1999)) as

$$x(t) = \int_0^t G_3(t-\tau)f(\tau)d\tau,$$

with the discussed homogeneous initial condition. For unit step function (when $f(t) = u(t)$), the solution may be obtained as

$$x(t) = \frac{u(t)}{m}\sum_{r=0}^{\infty}\frac{(-1)^r}{r!}\left(\frac{k}{m}\right)^r t^{2(r+1)}E_{3/2,r/2+3}^r\left(\frac{-c}{m}t^{3/2}\right) \tag{A.14}$$

and for unit impulse function (i.e., when $f(t) = \delta(t)$), the solution may be computed as

$$x(t) = \frac{1}{m} \sum_{r=0}^{\infty} \frac{(-1)^r}{r!} \left(\frac{k}{m}\right)^r t^{2r+1} E_{3/2, r/2+2}^r \left(\frac{-c}{m} t^{3/2}\right). \tag{A.15}$$

Now, one may see that the analytical solutions obtained for both the cases (unit step and impulse function) in Eqs. (A.14) and (A.15) are exactly the same as the solution obtained by HPM given in Eqs. (A.10) and (A.13), respectively.

In Eqs. (A.10), (A.13)–(A.15), $E_{\lambda,\mu}^r(y)$ is called the Mittag-Leffler function of two parameters γ and μ, where

$$E_{\gamma,\mu}^r(y) \equiv \frac{d^r}{dy^r} E_{\gamma,\mu}(y)$$

$$= \sum_{j=0}^{\infty} \frac{(j+r)! y^j}{j! \Gamma(\gamma j + \gamma r + \mu)}, \quad (r = 0, 1, 2, \ldots).$$

For unit step response $\gamma = 3/2$, $\mu = (r/2) + 3$ and for impulse response $\gamma = 3/2$, $\mu = r/2 + (2)$.

Here, HPM has successfully been applied to the solution of a fractionally damped viscoelastic system, where the fractional derivative is considered as of order 1/2. The unit step and impulse response functions with initial conditions are chosen to illustrate the proposed method. It is interesting to note that the results obtained by the present method exactly match with the solution obtained by Podlubny (1999), Suarez and Shokooh (1997), and Yuan and Agrawal (2002) in special cases.

A.2   FRACTIONALLY DAMPED BEAM (PROBLEM 2)

Let us now consider a crisp fractional differential equation that describes the dynamics of viscoelastic beam with damping

$$\rho A \frac{\partial^2 v}{\partial t^2} + c \frac{\partial^\alpha v}{\partial t^\alpha} + EI \frac{\partial^4 v}{\partial x^4} = F(x, t), \tag{A.16}$$

where ρ, A, c, E, and I represent the mass density, cross-sectional area, damping coefficient per unit length, Young's modulus of elasticity, and moment of inertia of the beam, respectively. $F(x, t)$ is the externally applied force and $v(x, t)$ is the transverse displacement. $\partial^\alpha/\partial t^\alpha$ is the fractional derivative of order $\alpha \in (0, 1)$ of the displacement function $v(x, t)$. We consider the initial conditions as $v(x, 0) = 0$ and $\dot{v}(x, 0) = 0$. The homogeneous initial conditions are taken here to compare the present solution with the solution of Zu-Feng and Xiao-Yan (2007). Equation (A.16) can be written as

$$\frac{\partial^2 v}{\partial t^2} + \frac{c}{\rho A} \frac{\partial^\alpha v}{\partial t^\alpha} + \frac{EI}{\rho A} \frac{\partial^4 v}{\partial x^4} = \frac{F(x, t)}{\rho A}. \tag{A.17}$$

According to HPM, one may construct a simple homotopy for an embedding parameter $p \in [0, 1]$ as follows:

$$(1-p)\frac{\partial^2 v}{\partial t^2} + p\left[\frac{\partial^2 v}{\partial t^2} + \frac{c}{\rho A}\frac{\partial^\alpha v}{\partial t^\alpha} + \frac{EI}{\rho A}\frac{\partial^4 v}{\partial x^4} - \frac{F(x,t)}{\rho A}\right] = 0, \qquad (A.18)$$

or

$$\frac{\partial^2 v}{\partial t^2} + p\left[\frac{c}{\rho A}\frac{\partial^\alpha v}{\partial t^\alpha} + \frac{EI}{\rho A}\frac{\partial^4 v}{\partial x^4} - \frac{F(x,t)}{\rho A}\right] = 0. \qquad (A.19)$$

Here, p is considered as a small homotopy parameter $0 \leq p \leq 1$. For $p = 0$, Eqs. (A.18) and (A.19) become a linear equation, that is, $\partial^2 v/\partial t^2 = 0$, which is easy to solve. For $p = 1$, Eqs. (A.18) and (A.19) turn out to be the same as the original equation (A.16) or (A.17). This is called deformation in topology. $\partial^2 v/\partial t^2$ and $\frac{c}{\rho A}\frac{\partial^\alpha v}{\partial t^\alpha} + \frac{EI}{\rho A}\frac{\partial^4 v}{\partial x^4} - \frac{F(x,t)}{\rho A}$ are called homotopic.

Assume the solution of Eq. (A.18) or (A.19) as a power series in p as

$$v(x, t) = v_0(x, t) + pv_1(x, t) + p^2 v_2(x, t) + p^3 v_3(x, t) + \cdots, \qquad (A.20)$$

where $v_i(x, t)$ for $i = 0, 1, 2, \ldots$ are functions yet to be determined. Substituting Eq. (A.20) in Eq. (A.18) or (A.19), and equating the terms with the identical power of p, we can obtain a series of equations of the form

$$p^0 : \frac{\partial^2 v_0}{\partial t^2} = 0, \qquad (A.21)$$

$$p^1 : \frac{\partial^2 v_1}{\partial t^2} + \frac{c}{\rho A}\frac{\partial^\alpha v_0}{\partial t^\alpha} + \frac{EI}{\rho A}\frac{\partial^4 v_0}{\partial x^4} - \frac{F(x,t)}{\rho A} = 0, \qquad (A.22)$$

$$p^2 : \frac{\partial^2 v_2}{\partial t^2} + \frac{c}{\rho A}\frac{\partial^\alpha v_1}{\partial t^\alpha} + \frac{EI}{\rho A}\frac{\partial^4 v_1}{\partial x^4} = 0, \qquad (A.23)$$

$$p^3 : \frac{\partial^2 v_3}{\partial t^2} + \frac{c}{\rho A}\frac{\partial^\lambda v_2}{\partial t^\lambda} + \frac{EI}{\rho A}\frac{\partial^4 v_2}{\partial x^4} = 0, \qquad (A.24)$$

$$p^4 : \frac{\partial^2 v_4}{\partial t^2} + \frac{c}{\rho A}\frac{\partial^\alpha v_3}{\partial t^\alpha} + \frac{EI}{\rho A}\frac{\partial^4 v_3}{\partial x^4} = 0, \qquad (A.25)$$

and so on.

Choosing initial approximation $v_0(x, 0) = 0$ and applying the operator L_{tt}^{-1} (which is the inverse of the operator $L_{tt} = \partial^2/\partial t^2$) on both sides of each Eqs. (A.21)–(A.25), one may obtain the following equations:

$$v_0(x, t) = 0, \qquad (A.26)$$

$$v_1(x, t) = L_{tt}^{-1}\left(-\frac{c}{\rho A}\frac{\partial^\alpha v_0}{\partial t^\alpha} - \frac{EI}{\rho A}\frac{\partial^4 v_0}{\partial x^4} + \frac{F(x,t)}{\rho A}\right), \qquad (A.27)$$

$$v_2(x, t) = L_{tt}^{-1} \left(-\frac{c}{\rho A} \frac{\partial^\alpha v_1}{\partial t^\alpha} - \frac{EI}{\rho A} \frac{\partial^4 v_1}{\partial x^4} \right), \tag{A.28}$$

$$v_3(x, t) = L_{tt}^{-1} \left(-\frac{c}{\rho A} \frac{\partial^\alpha v_2}{\partial t^\alpha} - \frac{EI}{\rho A} \frac{\partial^4 v_2}{\partial x^4} \right), \tag{A.29}$$

$$v_4(x, t) = L_{tt}^{-1} \left(-\frac{c}{\rho A} \frac{\partial^\alpha v_3}{\partial t^\alpha} - \frac{EI}{\rho A} \frac{\partial^4 v_3}{\partial x^4} \right), \tag{A.30}$$

and so on. Substituting these terms in Eq. (A.20) with, $p \rightarrow 1$ one may get the approximate solution of Eq. (A.16) as follows:

$$v(x, t) = v_0(x, t) + v_1(x, t) + v_2(x, t) + v_3(x, t) + v_4(x, t) + \cdots.$$

The solution series converge very rapidly He (1999, 2000). The rapid convergence means that only few terms are required to get the approximate solutions.

A.2.1 Response Analysis

In this section, response analysis has been made with respect to unit step and impulse loads as follows. Similarly to Zu-Feng and Xiao-Yan (2007), the external applied force defined by $F(x, t)$ has been considered as

$$F(x, t) = f(x)g(t),$$

where, $f(x)$ is a specified space-dependent deterministic function, and $g(t)$ is time-dependent process.

A.2.1.1 Unit Step Response The unit step load has been considered of the form $g(t) = Bu(t)$, where $u(t)$ is the Heaviside function and B is a constant. By using HPM, we have

$$v_0(x, t) = 0, \tag{A.31}$$

$$v_1(x, t) = \frac{fBt^2}{2\rho A}, \tag{A.32}$$

$$v_2(x, t) = -\frac{fBc}{\rho^2 A^2} \frac{t^{4-\alpha}}{\Gamma(5-\alpha)} - \frac{EIBf^4}{\rho^2 A^2} \frac{t^4}{\Gamma(5)}, \tag{A.33}$$

$$v_3(x, t) = \frac{fBc^2}{\rho^3 A^3} \frac{t^{6-2\alpha}}{\Gamma(7-2\alpha)} + \frac{2EIBcf^4}{\rho^3 A^3} \frac{t^{6-\alpha}}{\Gamma(7-\alpha)} + \frac{E^2I^2Bf^8}{\rho^3 A^3} \frac{t^6}{\Gamma(7)}, \tag{A.34}$$

$$v_4(x, t) = -\frac{fBc^3}{\rho^4 A^4} \frac{t^{8-3\alpha}}{\Gamma(9-3\alpha)} - \frac{3EIBc^2f^4}{\rho^4 A^4} \frac{t^{8-2\alpha}}{\Gamma(9-2\alpha)}$$
$$- \frac{3E^2I^2Bcf^8}{\rho^4 A^4} \frac{t^{8-\alpha}}{\Gamma(9-\alpha)} - \frac{3E^3I^3Bf^{12}}{\rho^4 A^4} \frac{t^8}{\Gamma(9)}, \tag{A.35}$$

and so on, where $f^{(i)} = \partial^i f / \partial x^i$.

Therefore, the solution can be written in the general form as

$$v(x,t) = \frac{B}{\rho A} \sum_{n=0}^{\infty} \frac{(-1)^n}{n!} \left(\frac{EI}{\rho A}\right)^n f^{(4n)} t^{2(n+1)} \sum_{j=0}^{\infty} \left(\frac{-c}{\rho A}\right)^j \frac{(j+n)! t^{(2-\alpha)j}}{j! \Gamma((2-\alpha)j + 2n + 3)}.$$

(A.36)

A.2.1.2 Unit Impulse Function Response

Unit impulsive load has been taken as $B = 1$ and $g(t) = \delta(t)$, where $\delta(t)$ is the unit impulse function. Again by using HPM, we have

$$v_0(x,t) = 0,$$

(A.37)

$$v_1(x,t) = \frac{ft}{\rho A},$$

(A.38)

$$v_2(x,t) = -\frac{fc}{\rho^2 A^2} \frac{t^{3-\alpha}}{\Gamma(4-\alpha)} - \frac{EI f^4}{\rho^2 A^2} \frac{t^3}{\Gamma(4)},$$

(A.39)

$$v_3(x,t) = \frac{fc^2}{\rho^3 A^3} \frac{t^{5-2\alpha}}{\Gamma(6-2\alpha)} + \frac{2EI c f^4}{\rho^3 A^3} \frac{t^{5-\alpha}}{\Gamma(6-\alpha)} + \frac{E^2 I^2 f^8}{\rho^3 A^3} \frac{t^5}{\Gamma(6)},$$

(A.40)

and so on, where $f^{(i)} = \partial^i f / \partial x^i$. Hence, the solution can be written in general form as

$$v(x,t) = \frac{1}{\rho A} \sum_{n=0}^{\infty} \frac{(-1)^n}{n!} \left(\frac{EI}{\rho A}\right)^n f^{(4n)} t^{2n+1} \sum_{j=0}^{\infty} \left(\frac{-c}{\rho A}\right)^j \frac{(j+n)! t^{(2-\alpha)j}}{j! \Gamma((2-\alpha)j + 2n + 2)}.$$

(A.41)

A.2.2 Numerical Results

In order to show the responses in a precise way, some numerical results are presented here. Equations (A.36) and (A.41) provide the desired expressions for the considered loading condition. As we have considered a simply supported beam, one may have the expression for the force distribution as

$$f(x) = \sin\left(\frac{\pi x}{L}\right).$$

The numerical computation has been done by truncating the infinite series (A.36) and (A.41) to a finite number of terms. Let us denote c/m and $EI/\rho A$ as $2\eta\omega^{3/2}$ and ω^2, respectively, where ω is the natural frequency and η is the damping ratio. The values of the parameters are taken as $\rho A = 1$, $L = \pi$, and $m = 1$.

A.2.2.1 Case Studies for Unit Step Response

Obtained results for various parameters are depicted in Figs. A.1 and A.2. Figure A.1 illustrates the effect of displacement against time for different values of $\alpha = (0.3, 0.6, 0.9)$. Figure A.1(a) and (b) presents the plot for $\omega = 5$ and $10\,\text{rad/s}$, respectively, with $x = 1/2$ and $\eta = 0.05$. A similar simulation has been done with damping ratio $\eta = 0.1$ and the results are depicted in Fig. A.2. Figure A.2(a) and (b) shows the plot for $\omega = 5$ and $10\,\text{rad/s}$, respectively.

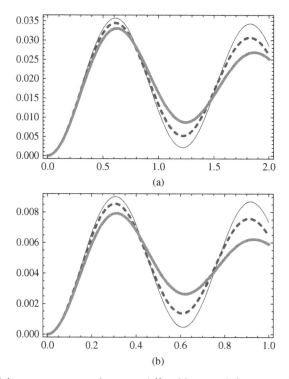

Figure A.1 Unit step responses along $x = 1/2$ with natural frequency (a) $\omega = 5\,\text{rad}/\text{s}$, (b) $\omega = 10\,\text{rad}/\text{s}$, and damping ratio $\eta = 0.05$

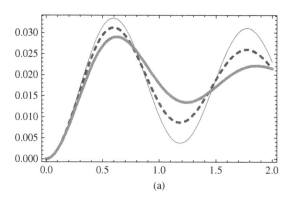

Figure A.2 Unit step responses along $x = 1/2$ with natural frequency (a) $\omega = 5\,\text{rad}/\text{s}$, (b) $\omega = 10\,\text{rad}/\text{s}$, and damping ratio $\eta = 0.1$

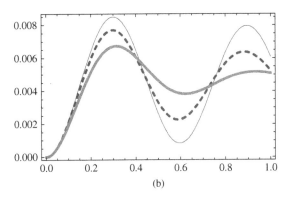

(b)

Figure A.2 (*Continued*)

It is interesting to note from Figs. A.1 and A.2 that if we increase the order of the fractional derivative α, the beam suffers less oscillation. It means that the beam suffers more oscillations for smaller value of α (solid thin line: $\alpha = 0.3$, dotted: $\alpha = 0.6$ and solid thick: $\alpha = 0.9$). It may be noted that in all figures in this appendix, the same notation for α has been followed.

A.2.2.2   Case Studies for Impulse Response   Equation (A.41) provides the desired expressions for the considered loading condition. In order to show the responses more clearly, some numerical results are presented in this section. Obtained results for this have been incorporated in Figs. A.3 and A.4. Figure A.3 illustrates the effect of displacement on time for various values of $\alpha = (0.3, 0.6, 0.9)$. In this computation, x and η are taken as 1/2 and 0.05, respectively. Figure A.3(a) and (b) presents the plot for $\omega = 5$ and $10\,\text{rad/s}$ respectively. Similar simulation has been done for damping ratio $\eta = 0.1$ and obtained results are depicted in Fig. A.4.

HPM has successfully been applied to the solution of a fractionally damped viscoelastic beam as well. The unit step and impulse response functions with crisp homogeneous initial conditions are chosen to illustrate the proposed method. The obtained results are compared with the analytical solution of Zu-Feng and Xiao-Yan (2007) and those are found to be in good agreement.

Readers may find it useful reading the method of handling the crisp fractional differential equation as given in this appendix for a particular type of example problems. This way it will be easier to understand the solution methods for the fuzzy fractional differential equations.

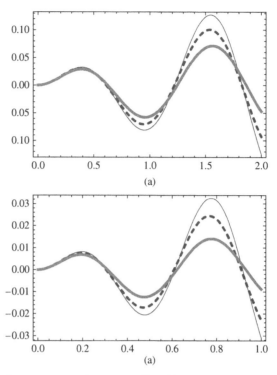

Figure A.3 Impulse responses along $x = 1/2$ with natural frequency (a) $\omega = 5\,\text{rad/s}$ (b) $\omega = 10\,\text{rad/s}$, and damping ratio $\eta = 0.05$

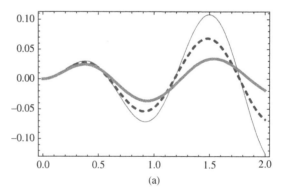

Figure A.4 Impulse responses along $x = 1/2$ with natural frequency (a) $\omega = 5\,\text{rad/s}$ (b) $\omega = 10\,\text{rad/s}$, and damping ratio $\eta = 0.1$

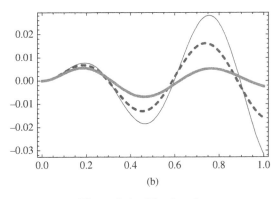

(b)

Figure A.4   (*Continued*)

BIBLIOGRAPHY

Enelund M, Lesieutre GA. Damping Described by fading memory-analysis and application to fractional derivative Models. *Int J Solids Struct* 1999;**36**:939–970.

Gaul L, Klein P, Kemple S. Impulse response function of an oscillator with fractional derivative in damping description. *Mech Res Commun* 1989;**16**:4447–4472.

He JH. Homotopy perturbation technique. *Comput Methods Appl Mech Eng* 1999;**178**: 257–262.

He JH. A coupling method of homotopy technique and a perturbation technique for nonlinear problems. *Int J Non-linear Mech* 2000;**35**:37–43.

He JH. Homotopy perturbation method for bifurcation of nonlinear problems. *Int J Nonlinear Sci Numer Simul* 2005;**6**:207–208.

Mohyud-Din ST, Yıldırım A. An algorithm for solving the fractional vibration equation. *Comput Math Model* 2012;**23**:228–236.

Momani S, Odibat Z, Alawneh A. Variational iteration method for solving the space- and time-fractional KdV equation. *Numer Methods Part Differ Eq* 2008;**24**:262–271.

Podlubny I. *Fractional Differential Equations*. New York, NY: Academic Press; 1999.

Suarez LE, Shokooh A. An eigenvector expansion method for the solution of motion containing fractional derivatives. *J Appl Mech* 1997;**64**:629–635.

Wang, Q., 2007, "Homotopy perturbation method for fractional KdV equation," *Appl Math Comput*, **190**, pp. 1795-1802.

Yıldırım A. An algorithm for solving the fractional nonlinear Schrödinger equation by means of the homotopy perturbation method. *Int J Nonlinear Sci Numer Simul* 2009;**10**:445–451.

Yuan L, Agrawal OP. A numerical scheme for dynamic systems containing fractional derivatives. *J Vib Acoust* 2002;**124**:321–324.

Zu-Feng L, Xiao-Yan T. Analytical solution of fractionally damped beam by Adomian decomposition method. *Appl Math Mech* 2007;**28**:219–228.

INDEX

Fuzzy Arbitrary Order System: Fuzzy Fractional Differential Equations and Applications, First Edition.
Snehashish Chakraverty, Smita Tapaswini, and Diptiranjan Behera.
© 2016 John Wiley & Sons, Inc. Published 2016 by John Wiley & Sons, Inc.